FULL SPEED

in

REVERSE

*Awakening Self-Worth,
Happiness, and Purpose*

By Amy Melvin

FULL SPEED IN REVERSE
© Copyright <<2023>> AMY MELVIN

ISBN: 979-8-89109-680-6 - paperback
ISBN: 979-8-89109-681-3 - ebook

For more information, email the author at happyamy75@icloud.com.

Join me at www.HappyAmy.net!

DEDICATION

First, this book is dedicated to my sons.
Everything I do is for you...

Second, this book is dedicated to the "uns"—anyone who feels unloved, unimportant, unworthy, unusual, and unwanted. May you never feel insignificant again. You are here for a reason. Realize happiness by finding the most important thing missing self-worth. After that you'll find self-love, and your purpose; right where it's always been...inside yourself. May the only "un" you ever feel again is unlimited...

TABLE OF CONTENTS

PREFACE

I thought
I could move out with our boys,
take a little time,
file for divorce,
move on.
But then,
a knock on the door.
I check the time—it's really late.
I am in bed.
Should I ignore it?
I hear the knock again.
I walk out of my bedroom,
down the hall,
and turn the corner
to the front door.
I look out the small rectangular window.
It's pitch black out.
I see two men—two police officers.
Why are they standing at my front door?
Why did they knock?
Why didn't they ring the doorbell?
I open the white door,
then the screen door,

and step outside.

It's chilly.

I wrap my arms around myself.

Are you Amy Melvin?

Yes.

Is your husband Shannon Melvin?

Yes.

Oh God. What happened?

There's been an accident.

Is it bad?

Yes.

Oh God. How bad?

Is he dead?

Yes.

I bend over.

I can't breathe…

All I hear is my heartbeat in my ears.

My stomach feels sick.

I crouch down.

Can

Not

Breathe.

   Was he on his motorcycle?

   No, he was in a Suburban.

   Is she dead too?

   The officers exchange a look.

   She's being flown to Marshfield with very serious injuries.

   They hesitate before asking their next question.

Do you know her name?

Yes, I know her name.

I provide her information;
one of the officers writes it down.

Is there anyone in the house with you?

Yes, my two sons.

How old are they?

Six and eight.

Is there someone you can call to be with you right now?

My dad. He lives right down the road.

They wait while I call him.

Pop?

Aim, you ok?

No. Shannon's been in a car accident. He died, Pop.

Oh my God. I'm on my way.

I couldn't cry, couldn't breathe. I begged for time to speed up, but it was standing still. How could the seconds take so long? I wanted to be past this moment. My heart was beating so fast. This could not be happening. I didn't feel like me. Nothing felt right. How could any of this be true?

The woman was his girlfriend. She just began to work for Shannon. He started a new business selling accessible products from Europe and met her on a business trip to Oregon in November, just ten months ago. I liked her. She was beautiful and sweet. For the last ten months I welcomed her. She ate with us. I never thought, never imagined. I trusted him with my life. My husband wouldn't...but he did.

I remember the night I realized she was his. In June, she was invited to have supper with us. We all ate and drank some wine. After I put our sons to bed, they kept drinking. All of a sudden, Shannon got up and left. I was confused. I didn't know where he was going. Then she stood up and said she was sorry. She walked out of my house and got in his truck with him. I watched it from my kitchen window. She put her face into her hands; leaned over and began to cry. She looked up and said, "Just go." He backed out. I watched my husband leave; with her by his side. My heart broke. I made it to the bottom of the basement stairs and sobbed as quietly as I could, so the boys wouldn't hear me.

Not long after, I moved out of the big house with our boys. I moved into my mom's house. After my mom died, Shannon bought her house as office space. Four years later, his business grew too large and it was vacant. I told him I was moving out and wanted to buy my mom's house back from him. He asked me to wait to move until he got home from a business trip. I agreed, but I couldn't wait. I knew he would question every item I took. I didn't want to deal with his comments on top of how broken my heart was. I moved out with the help of my best friend the day before his return. I filed for divorce last Wednesday because I wanted to be done hurting, done being replaced, and done thinking about why. The damage was done; he'd chosen someone over me. I just wanted to be done, to move on.

Now this…

At last, Pop pulls up
and the police leave.
It's only been a couple of minutes,
but it feels like so much longer.
I close the door.
I look at Pop and hug him—
I feel lost.
I cry,
and then it begins…
Shannon's dead.
How could this be?
What am I going to do?
Oh my God, the boys…
I have to tell our sons
their daddy is gone,
that he died.
What are they going to think?
Are they going to think what I think?
That we weren't enough?
For him to live? To not drive drunk?
He always drank, but he never drove drunk.
Why now?
Oh, my babies…
This is going to crush their little hearts.
They are six and eight
And now they have no dad.
They're never going to get what they're supposed to get,
the things that boys should get from their dad.

What am I going to do now?
I have to raise them
alone.
I don't want to be a mom and dad.
The phone rings.
It's Shannon's sister;
she's on her way.
I have no idea what time it is.
I have no idea what time the cops came.
I don't want to go through this.
I want time to stop.
Give me a minute.
Let
Me
Breathe…
But time doesn't stop.
I fear the morning
I don't want it to come.
I don't want the boys to wake up.
I don't want to tell them.
I glance over at a window;
it's still dark.
I am relieved, yet in agony
as we wait
for the sun
to come up.
And it does. Morning always comes.
I've never dreaded to see my sons,

Their footsteps coming down the hall,
I sit in a room full of people
in silence.
My plan A is over.
You know Plan A.
The one where you fall in love,
get married, have kids,
and live happily ever after.
Yeah, that one.
It's over.
Plan B:
no warning
always forced…
By
Something
Terrible.
We are in Plan B.
Shit. Wait,
I'm in Plan B…me.
There's no "we" anymore.
It's just me and my sons.
Oh my God.
It hurts.
Everything is on me now.

My happiness was gone in an instant, replaced with this wreckage. It's as though someone took my beautiful artwork and crumpled it up right in front of my face; like a piece of trash, then long-shotting it across the room into a garbage can. Everything felt meaningless. I didn't want morning to come. I didn't want my sons to wake up. How was I supposed to tell them? I was grateful Pop was with me. So was Shannon's sister and my best friend, Bobbi. Together, we waited until the sun came up. But when the moment came, it would all be on me to reveal the horrible truth.

My boys woke on their own that morning, coming out of their room sleepy-eyed and wondering why it was sunny.

"Why didn't you wake us up for school?" they asked.

Noticing the other people sitting in our living room, they looked to me for answers. I sat on the floor and asked them to come sit in my lap, wrapping my arms around each of them. Their faces etched with concern. A tear slid down my cheek as I forced myself to utter the words I dreaded most. "Daddy was in an accident last night.

Their concern turned to confusion.

"Is he ok?" they questioned, as fear began to set in.

I said, "No. Daddy is in heaven."

My boys' hearts broke. Saddness washed over all of us. The room was thick with shock and disbelief, drenched in brokenness. No one knew what to think. No one knew what to do. I focused on my boys. How would this affect them? I knew how badly it felt for me. I wondered how they felt inside knowing that their dad was dead. Were they thinking that he would never pick them up for overnights or come to their games again? I imagined that a part of their lives ended that day. Something died in all of us in that room. This was so much worse than telling them about the divorce. That was a cake-walk. They just accepted it. They were calm and their lives weren't turned up-side down. This was nothing like that. It was so hard that my memories are faded now. That day seemed like it lasted a week. Once we all settled into the news, the doorbell began to ring; and it never stopped.

It seemed like everyone in our town of 2,400 showed up to pay their respects. It was the strangest feeling I ever felt. I have never felt worse and more loved at the same time. My heart began to skip beats. Every person who came offered us food, flowers, warmth, tears, and added to the sense of unknowing of what my future would be like. My heart beat skipped abnormally for months. Our sadness would never truly fade.

The next day, I went to the funeral home to plan Shannon's funeral. The funeral director, Tim, had a template with blank lines to write all the details. Every part of the funeral needed a name for the person who would play a role that day. Tim offered a generic template that could be used for the obituary, but I hate reading those kinds of tributes. I always find them impersonal and leaving me with more questions than answers. I wanted to write Shannon's story from my heart. Tim smiled warmly and nodded with approval. I began by naming the surviving family—most importantly, our sons. The next part was naming those who had gone before Shannon. My mom, Shannon's mom, and older family members; like his grandparents.

The next question Tim asked was which local pastor I wanted to speak at the service. I thought for a moment and knew there wasn't one. I said, "Shannon wasn't religious."

"We can skip it," Tim said, sensing my hesitation. "We can choose a pastor for you."

"No," I said. "Shannon wasn't religious. I will speak."

Tim looked at me with kind eyes and gently emphasized, "This is no ordinary task. It can be very difficult for a loved one to speak at a funeral."

For a second, I imagined being seated at the funeral with my sons, listening to a chosen pastor offering his take on this catastrophe; his interpretation of this person he knew nothing about. Even if a pastor could find the right words, I knew immediately it would be their words, made up of God and reasons that Shannon would despise. As if he was there in my mind, I knew there was no one better than myself to convince everyone that everything would be ok.

I was the only one who really knew Shannon and what needed to be said. This was not an event I could simply attend as a passive listener. The brain is amazing, because in split seconds, I also remembered our marriage vows. Death had parted us, but my marital obligation was not yet finished. I hadn't come this far to neglect my final responsibilities as his wife. I knew my sons needed to hear this from me.

I shook my head. "It has to be me."

"Grieving people don't usually do well in such a role," Tim continued, trying to reason with me. "The speaking part is best left to clergy."

"Write my name on the line," I insisted.

Tim did as I asked.

Speaking at the funeral was just the beginning of the many tasks I would face. There was one final task that I knew would be much worse, but I had to do something. I was scared and this is not a feeling I'm used to having.

As we finished the rest of the funeral questions, I made my last big request known.

"Tim, I need to say goodbye to my husband."

"Amy, choosing to be the speaker at Shannon's funeral is not something I recommend, but I *really* don't feel it's a good idea for you to see his body."

"I have to," I said, knowing exactly what I wanted (but in all honesty, I had no idea how badly this decision would hurt). Tim knew. So he said, "I will tell you in very specific detail what Shannon's body looks like," he said, "and then you can decide if you still wish to see him. Ok?"

I nodded. Shannon's sister was with me and said she would go in with me for support. I protested immediately. I was not going to let her go in with me. This was something I needed to do and I knew exactly why I needed to do it.

Tim proceeded to tell details of what the accident had done to Shannon. He left no brutal detail out and concluded by saying, "Keep in mind, hearing what Shannon looks like is nothing compared to actually seeing the damage."

He also reminded me that Shannon's body was going to be cremated and was not prepared for a viewing. I didn't care. I could not leave without saying goodbye. He gave one last warning that this was not a good idea and asked for a few minutes to clean the blood from Shannon's face. Tim returned a short time later to get me.

We walk through the halls of the funeral home,
Tim knew I wasn't prepared for this.
Stopping in front of a door, he turned to me.
"I'd like to go in with you, Amy," he offered.
"No, I must do this alone.
I have to say goodbye."
I knew if I didn't, I would always regret it.
I knew I would never be able to erase what I was about to see,
but I had to.
Tim nodded and motioned toward the door,
"You don't have to do this."
I ignored all the reasons I shouldn't.
I grabbed the handle,
closed my eyes,
took a breath,
and put my head down.
I walked in and shut the door.
I closed my eyes.
I was too scared to open them.
The room was dark and quiet,
I opened my eyes to slits, looking down
letting in only what I could see past my eyelashes
I could see one small light

shining to my right.
I lifted my gaze as slowly as I could,
and saw Shannon,
lying on a cold metal table,
a white sheet was covering his body,
but not his head.
I focused on the end of the table
where his feet were.
I moved closer
until I was standing right next to him.
He's dead.
I'm standing next to his body,
I will never see him alive again.
I wasn't afraid, but remembered
the details Tim described.
I started to dread.
I dreaded
seeing
everything
too
fast.
My stomach aches.
I can hear my heart beating in my ears.
I force myself to breathe
and look at his feet.
Under the white sheet,
I started to walk toward his head,
as my gaze reaches his chin.

I see his goatee.
I look under his neck and there is a block of foam,
supporting his neck
where it is broken.
I look at his cheeks,
And then his eyes—they were closed.
But I notice
at the top of his left eyelid,
on the rim
where his eyelashes are,
there is a tiny horizontal cut.
How gentle that cut is, I think.
How did that cut happen?
My gaze rises above his eyebrows
where he always wore a bandanna.
It
Is
Bad.
So bad. Tim was right. I didn't want to see this.
I know instantly I will never forget this.
I squeeze my eyes shut and I'm overwhelmed for what he
did to himself.
I stand there a moment,
then I kiss his cheek three times—
once for myself, and once for each of our sons.
Goodbye, Shannon.
I turn and walk away...
I'll never see him again.

I will never forget what I saw that day. I will never regret it either. I had to see him. I was not only obligated to say goodbye; I needed to see him to have an understanding. Shannon was someone who lived his life on purpose. He was like me, in that, he pushed his life with severe intent. Seeing him this damaged gave me closure. My eyes had to witness what he did: to our marriage, to our love, to our sons, to his life, and to himself. The choices Shannon made took his life away from us. He was dead and didn't even care that he forced me into being the only parent our children had left. He didn't care that he made me a widow. And he did it while sitting beside another woman. I knew my eyes had to see what he had done, I had to witness the devastation firsthand. Remembering seeing the brutality of what happened to his body would help me control the sadness and hurt that I knew would come once the shock wore off.

I was numb. I knew later, when the pain set in; somehow, I would blame myself. I know this feeling because it happened when my mom died. I thought I was ready to accept her death, but the minute she took her last breath, I could only remember the things I could have done better. I couldn't remember anything bad she ever did. I knew it would happen again with Shannon. So, I forced myself to see his broken body—the hard evidence of his last choices—to remind me that I didn't do this. This was not my fault. *He* chose this. *He* pushed his life to this degree, and I was not to blame.

I walked out of the funeral home with my heart broken open by and death digging deeply into my being. We had only been separated for three months leading up to today. Three months ago, when I moved out, I was relieved. Instead of being overwhelmed by being replaced by another woman; I decided to look forward to being single and feeling happy to be free. I focused on the positives of finally living without Shannon's excessive drinking every night and sleeping apart from a partner who snored excessively from being drunk. Today, I could barely focus on anything. I could not mask how devastated I felt being replaced by another woman. I couldn't shake the unsettled dissatisfaction of knowing the marriage unraveling was such a minor detail. Death had swept over me, and I felt like a robot, an empty shell. It felt like everything was just happening to me.

Unfortunately, we were still figuring out the details of our separation when Shannon died, which meant my name was still on the mortgage at the big house. When I moved out with my boys, I moved into a home that my mom built. When she died, Shannon's business bought it. We were going to finalize me purchasing it back from him when we went through the divorce. Now, I was living in a home I didn't own. I absolutely knew I didn't want to move back to the big house we had shared for so many years. My best friend Bobbi reminded me that I didn't have a choice. She knew what I had to do and with kindness, suggested; "Being back in the home they know best will help the boys heal." I didn't have any other options, so we prepared to move back to the big house the next morning.

The next day came, and we were just about ready to leave my mom's beautiful little home when the severity of everything washed over me and I broke down in my mom's empty closet. Immediately, my poor boys ran in and huddled over me.

Frightened, they asked, "Momma, are you ok? What can we do?"

I was sobbing and empty. They had to be so scared. And instantly, as if a light came on something instantly healed me. I realized I was now *all* they had. They needed me to be strong.

I embraced my boys in a tight hug. "We are going to be fine," I reassured them. "All I need is you two."

And it was true. They were now mine and mine alone. We hugged each other for a lingering moment, and then gathered the last of our things to move back into the big house and begin to heal. My life went on autopilot that moment. It would be a few years before I realized my happiness again, but eventually, I would look back and recognize that I went full speed in reverse (FSIR) three times that week. I used my past to improve my future—a method that I would later employ time and time again.

INTRODUCTION

FULL SPEED IN REVERSE

The Full Speed in Reverse (FSIR) method utilizes an internal bank account you never knew you had, but it's been there since you were a child. In this internal account, you've deposited your disasters, your disappointments, and all your unmet desires. Unbeknownst to you, this account has been growing and collecting its worth in the form of strength for years, just waiting for you to put the devastation to good use. FSIR is the art of drawing from your past experiences and using the information and knowledge gained to enhance your present.

The first time I went FSIR was that day at the funeral home. I went back in my memories to the day Shannon and I said our wedding vows. In them were parts of Kahlil Gibran's "On Marriage" poem:

And stand together yet not too near together:
For the pillars of the temple stand apart,
And the oak tree and the cypress grow not in each other's
shadow.

I remembered truly feeling those words. Just as each pillar does its part to support the same structure, Shannon and I were two individual parts of our marriage that made up a whole—equal but set apart. We were like two separate trees, neither able to grow without the proper space to do so. If trees grow too closely, one will lack sunshine and the other will eventually overtake it.

Remembering our vows allowed me to not only take on the task of speaking at Shannon's funeral, but it also gave me clarity that Shannon *deserved* this one last support from me. He was not religious, and for me to allow a pastor who didn't know Shannon to speak about his life was not the right thing to do—I would have lived with that regret forever. Choosing to speak on our behalf freed me from ever feeling guilty about his death. None of this was my fault and I did everything I could possibly do.

Going FSIR means realizing that the easy road is one you must choose *not* to take. Even though no one would have blamed me for not speaking eloquently about the man who had just broken my heart, I chose to support my husband one last time. I chose to be kind. I remembered all the strength I'd deposited in my internal bank account. I could not trust our message in someone else's hands. The true story needed to be told by me. I spoke because I was the only one who could do it justice. I did it so I wouldn't have any regrets. Remembering our vows and using the strength I had built up internally; I stepped up, moved forward, and began to take steps toward happiness.

The second time I went FSIR was when I said goodbye to Shannon. Even though I knew it would be difficult to see his body, I needed to physically see the hard evidence of Shannon's final choices in order to free myself from any guilt or self-doubt that might come up later. And the third time, was breaking down in my mom's closet. Seeing the panic in my sons' faces as *they* tried to help *me* through this difficult time reminded me that I was their only parent now. They needed me more than I needed to surrender to my pain. I had to be strong and remain a consistent and reliable foundation for them to stand upon.

In each of these moments, I was left with two options: give up my power or take responsibility. This is where your internal bank account will pay you back with interest. Going FSIR allowed me to pick myself up and find my way forward having no regrets. Even though no one would have ever held anything against me during such a turbulent chapter of my life, I would have known that I could have done more while simply choosing not to. My inner strength set the foundation for my boys to heal, and for me to heal too. FSIR gave me the ability to confidently take full responsibility.

When my sons needed me that day in my mom's closet, I remembered back to when I was four years old, and my mom moved out. I wanted nothing more than for her to come back and show me she loved me. Her ability to simply leave me behind caused me to feel unworthy. But this childhood wound also gave me the ability to understand what I wished to receive when I was a child, allowing me to provide that very thing for my own children. They needed the same thing I did—they needed their mom. FSIR made me realize that my mom was not there for me the way I wanted and needed, but I promised myself that things were different for my boys.

Going FSIR gave me the strength I needed to be there for my sons when they needed me most. It allowed me to stay and take the high road rather become a victim of Shannon's death. Even though I knew I didn't have all the answers, I was going to try my hardest to be a dependable foundation for them. No matter what I was going through on the inside, my boys would never wonder where I was or whether I loved them. I would always be right there within reach, ready to listen, support, encourage, and validate them.

FSIR: using your past to transform your future.

AFTERSHOCK

A couple months after Shannon died, I began to feel sorry for myself. I anticipated it would happen after the shock wore off. This time it was not about my boys; it was internal. I felt awful, devastated that my husband chose another woman over me. I was mad that he replaced me so willingly. I was angry that he'd died and left me alone to raise our boys. No,

I wasn't *just* angry; it was so much deeper than that. My happiness was suffering. I was sad, scared, confused, angry, hurt, grieving, and exhausted. On the outside, I was keeping it together, but on the inside, it hurt with an unrivaled immensity. I knew I couldn't stay with this feeling forever—I had to figure it out. I didn't want my boys to see me struggle and cause them to wonder how I felt about them.

I was so low and decided to try to find something in my past to relate to. I immediately remembered a conversation I'd had years ago with my dad, Pop. Pop once told me after my mom left us, he'd struggled for six months. One day, he realized if she didn't want him, he didn't want her. It was like a light went on and he simply made up his mind that he would move on. Pop went FSIR. He took what he had saved up and used it to move forward. I found strength in this and did the same thing. I remembered back to all the holidays we spent at my mom's house with her new husband. Pop was at every single one of them. We would wake up at my mom's house on Christmas morning and wait for Pop to get there to open gifts together. He shook my stepfather's hand, ate at the same table, and never once made anyone feel uncomfortable. The way Pop handled his divorce was one of the greatest gifts I ever received as a child. I stored that up in my internal bank account and was able to use it today.

Even though Shannon gave up on our marriage and moved on with someone else; I barely knew the other woman. What I did know was that she was smart, warm, and very beautiful. I wondered if she made Shannon happy.

I began to wonder why I didn't; but instead of focusing on all the what-ifs, I decided it was time for me to move on too. I admitted to myself that I would rather have had him alive and happy (even if it was with someone else) than dead. Everyone deserves to be loved. Shannon deserved it too. I let go of my right to be a victim, deciding instead to use everything I had inside of myself to move on. I forgave Shannon and when you know me, you'll know that I always go ten steps further and called the woman. She didn't answer, but I left her a voice message telling her I forgave her; and I wished her well. I told her she received a gift that night, by not dying in the accident. I hoped she would use her gift and do something wonderful with her life. I also realized that she might miss Shannon for the rest of her life. When I hung up, I immediately felt free. I knew I'd done the right thing—not for Shannon or his girlfriend, but for myself. My happiness was beginning to be restored.

Going FSIR allowed me to use the past to be honest with myself today. Being honest opened a path for the truth; not just the parts I *wanted* to see, but every aspect of the entire picture. Remembering Pop's story about his journey to his own healing opened my eyes to see there are thousands of people going through these types of situations. I wasn't the first person who didn't get what I wanted. By connecting to Pop and the other woman, I was finally able to accept that and find peace with it.

I call this awakening method, FSIR and it results in happiness being realized. It's stepping outside of yourself and observing yourself. It allows you to connect to others who have experienced similar things, but goes deeper because observing yourself allows you to be able to see your own life through a different lens. It allows you to analyze your situation truthfully and find the path forward using what you received from your past. Pop did it when he got divorced. And I was able to do it because of Pop's story and displays of strength over the years when I was little. I realized that I am responsible for leveling my internal scales and finding my own balance. It was a choice, and no one could have done it for me.

I opened the door to realizing my happiness by going FSIR. It's been almost ten years since Shannon died and the happiness I realized then, is still going strong today. FSIR provided a way for this to happen and continues to resonate. I have been able to show up in a positive way restoring happiness to my heart. It's brought me closer to my sons. It's allowed me to let go of how I wanted things to turn out and to accept them the way they are. Most things in life usually come down to two very clear options: accept or resist. Acceptance will always move you forward, allowing growth and healing to take place. It was the best decision I ever made. Resisting would have changed nothing.

The tools that follow will give you the power to learn to react and navigate differently through any circumstance that comes your way. By using the FSIR method you will learn how to show up for yourself first so you can show up for the people in your life who need you. You will gain strength building your internal support system. You will be able to release guilt, fear, worry, anxiety, longing, and all of life's burdens. I don't want you to go another day without knowing how to go FSIR. Happiness Realized begins now...

CHAPTER I

WHO ARE YOU NOW?

The first step to realizing your happiness is by increasing your self-worth. Self-worth must come before any self-help guidance can ever work. This is a pre-self-help book. If you've ever struggled with feeling worthy, you're not alone. This vital element is missing from so many people and can be changed easily by working on the inside. I want you to know that you alone are in charge of determining how much or how little you value yourself.

Begin by thinking about who you are right now? What do you love about yourself? What value do you bring to the world? What is your purpose? I know these are big questions, but this is an important first step in your journey to realizing happiness. I want you to think about all you have achieved. What are the pieces that you consider a major part of your identity? You may find it surprising to look at all the things you've done with your life.

I'll start by telling you about mine.

I am Amy Elizabeth Greene Melvin. I am from New Jersey. I am a confident, strong, and independent woman. I am a widow, a mother, a daughter, and a sister. I am very protective of myself and my sons. I am easily excitable and empathetic. I am a woman who wants to evolve and positively control her mind and her reactions. I believe anything is possible. I am an author and a writer, a devoted skier, and an experienced motorcycle rider. I have earned a bachelor's degree in psychology. I am unafraid of almost anything and am not concerned with what others think of me. I am a certified personal trainer and an exercise enthusiast. I am in love with my body. I am a pursuer of knowledge. I am someone who strives each day to become a better version of myself. I am skilled with meditation, drawing, and willpower. I adore music. I am a champion of people. I was born to inspire others. My goal is to change the world by inspiring people to realize their limitless potential. I am someone who loves change and who strives each day to observe myself. I long to be comfortable, being uncomfortable. I am elated to be my unique and brave self and am learning to find more happiness than I ever imagined each day.

Take a moment to write down who you are. Start with your full name and write down your skills, personality traits, passions, beliefs, goals, and other identifiable features.

I am _____

_____

_____

_____

_____

_____

_____

_____

_____

_____

_____

_____

_____

_____

_____

_____

_____

_____

_____

_____

_____

_____

_____

Now pause and read over what you wrote. How does your list make you feel? Are you proud of yourself? I am. Your list makes me proud! Everything you wrote is an accomplishment. Everything. It doesn't matter if you have one accomplishment or fifty, they are reasons for celebration and pride. What you have done is amazing. Believe it and feel who you are now.

Who you are now is an accomplishment. When I look back on what I wrote, each one of those things took me years to accomplish. I didn't start off my life loving my body or realizing that my purpose was to inspire people. I grew into every one of those things I became. Self-worth has everything to do with my list. Self-worth has everything to do with yours too. Without self-worth we wouldn't be where we are today. When you increase your self-worth, who you are will increase too.

THE POWER OF YET

Does your list make you proud? Are there things on your list that you worked hard for? Does your list shine light on things you *wished* were on your list? When you think of what you're not, it shadows who you are. Feeling disappointed in yourself because there are things you want to add to your list will get you no closer to where you want to be. There may be a ton of things you don't know how to start, much less finish. If you'd like to do more with your life, the first step is believing anything is possible. Next, be grateful for where you are now. Take a moment and realize that you got yourself here. Your list will change and grow along with you. Think about the hardest thing you've done or something you are skilled at today. Now think about where you started from and how far you've come.

Five years ago, I didn't have a bachelor's degree and I wasn't an author. I could look back and feel disappointment in all the things I could have accomplished before now. I could dwell on the fact that my list is small. I could be disappointed that there aren't a hundred more things on my list. I could hold it against myself and claim that my list isn't good enough. Or I could remember that my list isn't finished, yet. My accomplishments aren't complete, yet. I have not reached my fullest potential, yet.

Don't underestimate the power of the word, "yet." Yet is a word that brings possibility to anything you desire in your life. Just because you don't have your dreams completed today does not mean they are not possible. My list will grow, even though it hasn't yet. I use FSIR to remember back to the time when I didn't have my degree. There was a desire deep inside of me that wanted a higher education. I've always wanted my degree, but I didn't know how to accomplish it. I used everything I had from my past to strengthen the belief that I could accomplish what I wanted. Today, I have my degree. I will use FSIR to remember the inner strength I used to push myself. That strength allows me to keep setting and accomplishing my goals. I set a goal to write this book. I believed I could do it and so I did. My next goal is to add "motivational speaker" to my list. I feel it calling to me and even though I haven't accomplished it yet, I know I will.

Your list is going to change and grow, too. Use FSIR to remember what you've gone through and then realize your strength. Push past your doubts today realizing that you may not have accomplished everything on your list *yet*, but you will; if you believe that what's inside of you is all you need to get started. You are going to become what you've always dreamed of; I know you have it inside of yourself to get there. Just because it hasn't happened yet, does not mean it won't...

SELF-WORTH BEFORE SELF-HELP

When someone wants to make a change, the first place people often turn to is the self-help section. The self-help experts are stereotypical men, who are loud and passionate. They yell things like "Live up to your purpose!" "Work harder, do more, be more!" "Stop making excuses and just do it!" To be honest, I freaking love those guys. They are all fired up and yelling because they're absolutely sure they can change your life! And they're right, they will change your life; but there's one little thing you need *before* self-help books. The answer: self-worth. Those guys have tons of self-worth, it's what makes these gurus so confident. What are you missing that can make or break your transformation? Self-worth. You have to increase your self-worth because how you feel about yourself directly determines how far you are willing to go. So, before self-help can work, you need to believe that you are truly worth it. Otherwise, no amount of self-help is ever going to make you change anything about your life. Once your worth grows, you will feel more confident, and you will start to move mountains. If you continue increasing your self-worth, who you can become will be unlimited.

But self-worth can't be handed to you by a book, your mom, or drilled into you by a self-help expert. Your self-worth must be created within. This process begins by being honest about who you are *now*. Holding things against yourself won't do you any good. If you allow yourself to get stuck on where you aren't, you'll be no closer to your goal tomorrow. Always keep in mind that you're not done *yet*. You're just beginning to create yourself. Go FSIR and start at the beginning. Be grateful for your present self. Even if you want to be more. Even if you wish you could have done more by now. Realize that you are the greatest version of yourself exactly the way you are! Love yourself and all you have achieved. Then, take that your inner strength and set a new goal. Believe in your heart that you can reach your goal and know that your life has prepared you for whatever you wish to do.

Say this out loud:

I am grateful for who I am. I am grateful that I have survived every day of my life. I am worth it. I am worth every single thing I want to do, become, or create. I am a valuable human who is here for a reason. Everything I have, I am grateful for. Everything that has ever happened to me has made me who I am today. Even if some things weren't perfect, I am grateful for what I've learned from every experience I've ever had. I can accomplish anything my heart desires. I can accomplish any dream I have. I can move mountains. I am unlimited.

Increasing your self-worth begins with being grateful and appreciating who you are now. You will move ahead much quicker with gratitude than you will when you are disappointed in yourself. If you don't feel grateful yet, I am sorry you're not where you want to be. I'm sorry you've had bad days. I'm sorry you've been through tough times. I'm sorry if you've experienced loss. I truly am so sorry. Your internal bank account is brimming over with the strength that all those difficult things you've been through has created. If you had a meter on yourself, like a phone has a signal meter, your strength would be five bars full of life!

I urge you to remember that you're not done yet. Today is the first day of the rest of your life. Even if you've started this journey fifty times in the past, be proud that you can begin again – this time with a little more self-worth. Being grateful for who you are today is the difference that will make this the beginning that counts.

CHAPTER 2

DISCOVERING YOUR PURPOSE

I ncreasing self-worth will automatically increase self-love. The next discovery will be your purpose. Those three things make up the interlocked trinity that leads to realized happiness. They are all connected. If you're low on self-worth, your self-love is very likely to be low too. With both *selves* low, you may not have found your purpose...*yet*. The fact that you are reading this book right now means you recognize that you are worthy, even if your self-worth is very small. By building your self-worth, your self-love increases; when you feel good about yourself you will be more confident. The more confident you are, the more you will set out to do. The more you set out to do, the more you will accomplish. The more you accomplish, the bigger you will dream. The bigger your dreams, the closer you will come to finding your purpose. When you find and live out your purpose, your happiness will be realized.

I want to be very clear when I say that your self-worth might be low, but you have self-worth already. Everyone has it. It's the reason you wake up, go to work, get married, have children, and why you choose to do anything at all. If you didn't have any self-worth, your life would lack meaning. Some people have more self-worth than others because they discovered it during their childhood, and they were able to build upon it every day of their lives. Ideally, self-worth would have been instilled by your parents, but not all of us got our self-worth from our parents. Some children received worth from both parents. Other children may have only gotten it from one parent. And still others might have had a special person outside of their parents or even outside of their family, who instilled self-worth.

You might be one of those people who didn't have anyone in your life who felt proud of you. This is a unique kind of beginning because you probably began your life in survival mode. Survival mode almost always creates people who are either very reserved or very outgoing. You rarely accept (or believe) compliments, and you don't let others in easily. If you are a child of survival, you might process difficult things like the age where your survival began. Some things that trigger you may hold you back or keep you stuck. In survival mode it can take a long time to become self-aware. It's the reason children can only think of what they need. They haven't grown up to realize that there is a whole world of people who also need things. Make no mistake. The self-worth created by children of survival is a very special kind of self-worth. It goes so deep you may not know how to be proud of it.

If you don't currently feel your worth, do not worry. You have worth and you're not alone. So many people didn't get the gift of worth from a loved one and I'm sorry about that. I'm here today to tell you that you *should* have been raised by someone who was proud of you. You *deserved* to have a parent who loved you and who took pride in you. You deserved a happy childhood and an easy childhood. But even if you didn't have these things, the good news is your childhood does not have to define you any longer; nor does anyone around you. You oversee your self-worth, which means even if you don't feel much of it right now, it is in you; and you can increase it.

Go FSIR and start to believe that no matter how anyone feels about you, including yourself, you are an original. There is only one you, living on this earth right now. You were chosen to be alive. You have air in your lungs. You have a powerful mind. You have a spirit that is capable of literally anything. You deserve love and you are worth it. Most importantly, you are unique, different from anyone else, and one-of-a-kind. Every experience, hope, and dream you've ever had is yours. You are the only one who knows what you've been through, so begin to take credit for being the biggest part of your own life. Acknowledge your importance, feel your own love, and increase your self-worth.

SURVIVAL

Surviving is an art. It's never a choice and it's usually the result of a difficult childhood. It's the hardest way to grow up. If you are a child of survival, your childhood may have been overshadowed by things like uncontrollable siblings, disappointed parents, getting lost in relationships, engaging in intimacy too early, doubting yourself, raising yourself and/or having to raise your siblings, or being paralyzed by failures. If you fall in these categories, you might feel as though you aren't worth anything. You may have formed a protective shell that protects you from being hurt by others. Your protection may also be so thick that you have a hard time getting through things. You might have a hard time accepting compliments, feeling self-pride, and many other positive outside influences because you don't trust anyone.

The good news is that survival mode can be changed by looking at yourself from a new perspective. Go FSIR and begin to look at why you react to certain things or why you do the things you do. Becoming aware of your behaviors will help you open yourself to realizing that there are other ways of behaving. When you look at your own behaviors, you'll begin to see that they don't just affect you; they also affect other people. They are also just one way of reacting. Being open to your own outside perspective can shed light on trying to handle things differently in the future. Once you begin to see yourself, you might find that you soften to your own heart—maybe enough to encourage you to crack open your shell to new feelings of how worthy you really are.

It's true that receiving worth from your parents would have given you a wonderful boost, but if you missed out on that in your childhood, it doesn't make self-worth unattainable; it just makes it more *contained*. Remember, you built your own worth inside of you, so don't get weighed down with what other people didn't help you create. Maybe the people who raised you didn't have anyone who had pride in them either? Maybe they had to build their own self-worth and were too busy surviving their own childhoods to learn how to provide a better childhood for you? Your parents may have been raised by imperfect parents too? In fact, you could come from generations of people who are low on self-worth but remember; you have what you need. You can be the one to break the cycle. By going FSIR, realize it was *you* who got you through your childhood, and this is the very reason to look inside and increase your self-worth.

Forget about poor parenting and all the difficult things from your past and take pride in yourself for doing it on your own. Don't let those s-hackles and chains hold you back any longer. Your life may not have been easy but take pride in the fact that you made it through every second of your life on your own. You are amazing! Build upon your deep internal worth. Because this relationship you create with yourself is the most important relationship there is. You know *everything* there is to know about you. Every thought, feeling, and experience you've ever been through is yours. You've made it this far.

You would not be here if it wasn't for you. Every day, you've chosen to be alive. Every day, you wake up and keep moving forward. Every day, you might even show up for someone else who needs you. You are very important, so even if that feeling is small within you now, it's time to open your eyes and realize you hold tremendous worth. You are invaluable.

Never put your happiness on hold by continuing to let your past rule your life. When you increase your self-worth, you will begin to love yourself and a limitless you is born. With a strengthened self-worth, you will have the confidence to try new things, take risks, and go toward the future of your dreams, no matter what happened in your past.

DIFFERENT PERSPECTIVES

My mom left when I was four and my sister was six. As far as I knew, she had *chosen* to leave us behind. I grew up thinking my mom left because she didn't love us. Years later, my dad told me he was the one who'd said she couldn't take us with her. Regardless, she left my sister and me behind. Her seemingly selfish choice was hard on all of us. Pop did the best he could, but he had a very short temper, and he was stubborn. Even though he'd told my mom to leave us with him, he wasn't prepared to be a single parent. In his way of adapting to being our primary parent, Pop's parenting style was unconventional. We never had a bedtime because he didn't want to fight with us each night. Instead, he would let us fall asleep in his bed watching TV, and then he would carry us to our beds. He waited on us hand and foot, never making us help with dishes or cleaning the house. Every morning before school, he would make our beds and bring us cereal while we watched cartoons on the couch.

When my sister and I would go wild, running around and being obnoxious, Pop would come unglued as though all the avoided turbulence hit him at once. He would spank us and yell; we would cry. Later, he would come into our bedroom, sit on this little metal chair with a yellow plastic seat, and he would apologize. He said he was sorry for saying mean things and spanking us. He would plea with us to be better, and we would hug him and apologize and promise to be good girls. I always took those conversations to heart, but as my sister got a little older, she grew more out of control. In hindsight, I believe she blamed herself for my mom's decision to leave and she didn't know how to handle feeling like the one to blame. Seeking attention elsewhere, my sister turned into a promiscuous girl and her relationship with my dad began to spiral. With every passing day, it only got worse. They fought about her attitude. They fought about her loud music. They fought about her grades. They fought about the same things, over and over, for years. Even though my dad didn't have a lot of rules, as my sister got older, she continued to push the limits.

My dad's lack of rules likely stemmed from his own childhood. He didn't grow up with much parental guidance himself. Allowed total freedom to navigate Queens, New York from a very young age, his parents had just one rule for him to follow: stay out of trouble. So, he did. Even though he lacked guidance, he found his own self-worth through his independence. Able to spend his time the way he wanted; Pop chose to start working at a very young age. By making his own money, he never had to ask anyone for anything, including medical care. When his parents talked to him about making an appointment for the dentist around eleven or twelve years old, Pop told them that he had been going for a year with his own money. He was even allowed to quit school at age fifteen, opting to move to Florida to work with the Standardbred race horses. Having been on his own for so long, Dad had learned how to design his own life tools without his parents' help.

But one thing I wished my dad had gotten from his parents was my grandpa's patience. My grandpa was the kindest soul on earth. While my dad was blinded by anger, he blamed their troubles on my mom's faults. My grandpa recognized knew she was kind, unique, and outgoing, without an enemy in the world. Unable to look outside of himself to understand her, my dad was uncomfortable and embarrassed by my mom's personality. Whenever Dad would become wild and angry, my grandpa would come take Mom away and try to try to save her from Pop. I'm sure it was because he wanted to help her understand him. He would explain that my dad had been independent for so long, all he could see was himself. He couldn't see things from anyone else's perspective.

Some tools you get, some tools you create, and others are simply missing. Each of us received something, created something, and maybe missing something. For most of us, the missing link is just a connection. Because of their lack of understanding for each other, my mom and dad lost their connection. The same thing happened between my dad and my sister. He only saw his side, what he wanted to see. My sister was a child, so all she knew and cared about was her own perspective. Between Mom leaving and Dad always being unhappy with her, my sister soon learned to become what she heard she was. Diving into the influence of the people she hung out with, she would stay out late, lie about where she was and who she was with, and do things she shouldn't have. My dad couldn't trust her and it made him furious. He wanted her to do what he said, but she didn't care what he wanted. Every day, they did the same things that never worked, and the only thing that resulted was the deterioration of their relationship. Both tried to force their way upon the other; neither one was willing to bend.

Neither my dad nor my sister was willing to see things from the other's point of view. That's what was missing. Neither one took the time to understand the other. They couldn't step outside of themselves to see how their behavior affected each other. In reality, neither one of them would have had to change what they wanted, but in striving to understand each other, they could have made a connection that would have led to peace for them both.

I've learned the hard way that connections are generally what's missing from any failed relationship. When two people are connected, even if they don't see eye-to-eye, understanding each other always creates respect for the other person. Without this connection, my sister didn't get what she needed. My dad didn't either. Sadly, many of us don't. It was years before they created a new relationship with each other. My sister and Pop suffered from their failed connections. They were both held back by it. Not receiving what you need when you need it can hinder you from finding happiness and inhibits the ability to thrive.

Trust in the power to see something other than your own perspective. Understand that your choices usually affect others. I have used this knowledge to create a greater sense of understanding and have been able to create amazing connections in my life. Going FSIR, I remember back to my past and can utilize what was missing to create opportunities today with my own sons. I can make a positive turn toward better understanding others in my life which creates stronger connections and has allowed me to build my sons' self-worth. I have two of the strongest bonds with my sons than anyone I know. They not only come to me, they trust me; but I also go to them, and I trust them.

USE YOUR PAST

I know you've heard this before: your past is over and can have no effect on you, unless you let it. Most of us know that instead of allowing your past to rule over you, you can use your past for your benefit. But how? You can go FSIR and ask yourself if there's anything you can change about your past. Ask yourself what makes you unhappy about it? If there are things you can change, take action; and change what you can. If it's over and there are things you cannot change, accept where you are and move forward from here. Forgive yourself for your past, especially the things that can't be changed. Letting go is your first step toward loving yourself exactly the way you are today.

Do not be fooled—you are not defined by your past nor do you belong to your past. Where you were born, how you were raised, or anything else that happened in the past can only hold you back if you let it. Never allow it to be the reason you stop moving forward. Your past is always going to be yours, but being limited by it is a choice you make. Let go and begin to use your time wisely to increase your self-worth. It's first, about building a strong connection with yourself; after you do that, you can build connections with others. The bond you create with yourself will be the most important relationship you ever have. I grew up wanting a relationship with my mom and never got what I hoped for. Pop and my sister were so disconnected; but in that I've learned how to connect to myself. Go FSIR and use lessons from your past to improve your present. My connection with myself has given me the ability to realize that I've always had happiness, I just failed to realize it because I was too focused on what I was missing to see what I already had...

CONNECT WITH YOURSELF

Think of the first time you fell in love. Remember how good it felt? That person's call meant everything because it meant they were thinking of you. Their touch sent your skin tingling. Their attention felt amazing! That person choosing you meant everything because you were important to someone else. And then, it ended. That person stopped calling even though you longed for them. Your heart ached when you saw them with someone else. You were devastated. You were left to wonder, were you even worth anything without them?

Today might be the first time you're hearing this, but self-worth is the most important connection you have. It's your responsibility to grow and to nurture it. You were never supposed to give your worth away—not even a little bit of it. Your worth is *yours*. It belongs to you and only you. If you knew this before you fell in love, you would know to never give it away or look for it in someone else.

Think of your self-worth as a rare and valuable bank account that's totally unique to you. It's like money that you can't give to anyone else because the moment you give it away, it loses its value. When you rely on someone else to show you your worth, it's too much responsibility to bear.

The other person gets weighed down with the responsibility of your worth, so they let you go. It hurts so badly when the person you relied on walks away, but the cycle is doomed to repeat when you continue searching for your worth outside of yourself—somewhere it will never be.

Before you fall in love with anyone, you should *seriously* fallen for yourself first. If you had taken the time to connect with yourself instead of focusing your energy on someone else, you would have mastered your likes, dislikes, talents, passions, desires, feelings, purpose, and more. You would have been able to do the things you were passionate about because you would feel confident going toward those things. You would have chosen people who aligned with you. But the opposite usually happens. Most of us start dating so early, we don't take the necessary time to find out who we are, and we dive into someone else's life instead. Ten or twenty years later, you find you have a ton of alone time because you've got nothing in common with the person you chose. Or you can't even figure out how to pick the right person because you still don't know what you want or need. If you never focus on yourself, you will never increase your self-worth, this has the tendency to lead to a very unhappy and unfulfilled life.

When I was little, my situation allowed me to connect with myself. I didn't realize how important it was back then. Since Pop and my sister were so busy fighting, I spent a ton of time alone. I taught myself how to draw, discovering a love for art. I would also write and dream; and keep to myself. I spent hours connecting to who I am inside. I formed a deep relationship with myself. My parents expected very little from me, making it easy to get to know myself because I didn't have to please anyone else. I didn't date until I was much older because I didn't *long* for someone else; I had myself. I was my own best friend, and it was a blessing. The older I got, the less time I spent alone. As many do, I soon fell victim to my job, relationships, family life, getting hooked on television series, studying for college, getting wrapped up in social media, and other things. When I was young, there weren't cell phones or computers; the people in my life were all I had. Now, we are connected to our phones as though they are our hearts. Ask yourself if you may be under the influence of the need to be constantly entertained? When was the last time you spent time alone? I mean truly, alone?

If you want to connect with yourself, you need to spend time with yourself. If you've never done it before, you'll need to go into it with an open mind. You have to be willing to listen to yourself. If you've never listened to yourself, alone time can be magical. As an unfortunate side effect, it might also shed light on who you have not become. No one wants to be reminded of what they haven't achieved, so in the beginning of your alone time; if you begin to feel negative or weighed down, remember to be kind. Using the time to blame yourself for where you are not, will not get you to a better place. It also won't help you connect to yourself. In fact, this is the exact opposite of what's intended during this time of self-reflection and connection. I'm going to say it again: do not get caught up in who you're not. Remember, as long as you're breathing, you're not done *yet*.

Give yourself credit for coming this far. You are building a new relationship and beginning to love who you are right now. Use this connection with yourself as a chance to build self-worth and admiration on the inside. Let your alone time inspire you and help to understand where your feelings come from. Dive into what you struggle with and gently listen. If there are things you wished were different, write them down. You'll get there.

In your alone time, the goal is to connect with yourself and be open to what your heart tells you. Creating self-worth is the foundation for happiness, but it will require honesty. There are parts inside of you that no one else will ever see. You are the only one inside your head. Go inward and begin to change the way you feel about yourself. FSIR is going back and seeing where you came from and realizing, good or bad; you were the one who got yourself here today. Don't let the idea of being alone scare you. If it does, ease into it. Schedule an afternoon or morning to take a blanket and relax at the park or visit a museum. Focus on your good times and give yourself a break on the bad times. Think about who you are and what you want. Think about what's not working in your life and how to fix it, but don't be overwhelmed with changing everything at once. Instead, make plans to change what you can and daydream about the things you want to accomplish. Talk yourself *into* things instead of out of them and write your goals down to make them feel more real.

Take time for yourself and feel your own value. The confidence to change will only come from increasing your self-worth and knowing yourself deeply. That kind of knowledge will bring clarity to your passions, the connection you build will also help to reveal your purpose. So, accept this very important solo mission and give yourself the gift of time by focusing on being present and strengthening the connection to your current self.

Don't forget the importance of remaining focused on the here and now rather than dwelling on the *shoulds*. There are many things I wish I had done sooner in life, but I didn't. Staying stuck in the past will only overwhelm you and waste more of your time. Reminding yourself of what you haven't yet accomplished is unproductive and can cause happiness and self-worth to disappear. Practice increasing your self-worth by loving who you are currently, accepting your past, and being gentle about where you are today.

If you're unhappy with where you are, I need you to know that you can go further. It's never too late. If there's something you haven't accomplished yet, write it down and save it. But before you work toward something new, you need to first create pride in and build a connection with who you are *right now*. Spend some time alone. Find out what makes you come alive, what you've always wanted to do. Then go back and finish high school, get your college degree, get better grades, take harder classes, study something new, and become the person you've always dreamed of becoming, because life without passion isn't really living.

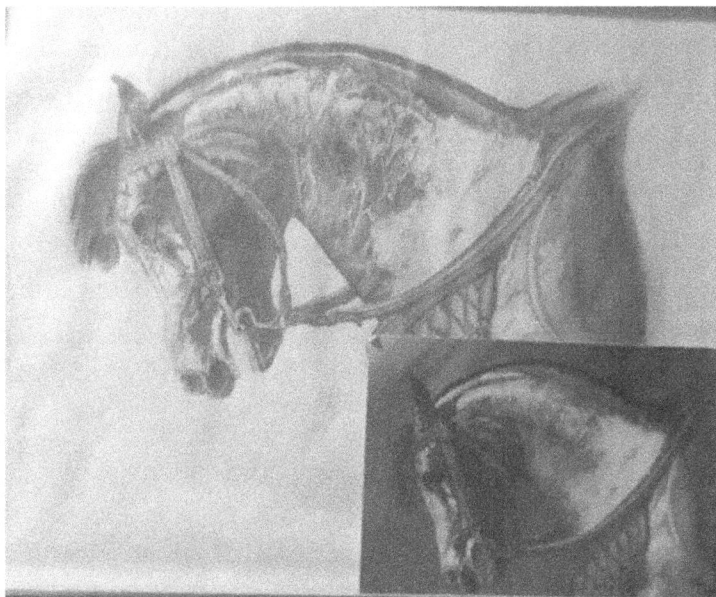

Figure 1: My passion for drawing continues to this day. I still spend hours teaching myself how to draw. Above is a pencil drawing I completed in 1993.

YOUR PURPOSE

My favorite question to ask is "What are you looking forward to?" You can tell so much about a person by how they answer this question. Some people will look up and wonder, unable to think of anything. It's very likely those people are not living their purpose, nor do they know what a limitless life is. And then there's the other kind of people—the confident and happy ones. When asked this question, their eyes light up with the possibilities. They smile, thinking about their life and what they're looking forward to, going into detail about upcoming celebrations, vacations, or milestones. They are living fully, sharing life and happiness in a big way with their loved ones. When I see someone light up about life, it's a gift of hope that there are people doing more than simply surviving. It inspires me to continue doing the same. Life is meant to be enjoyed. This starts with knowing how worthy you are, then you will be able to confidently pursue what you want.

So, what are you looking forward to? Write it down below.

_____

_____

_____

_____

_____

_____

_____

_____

_____

_____

_____

_____

_____

_____

_____

_____

If you're finding it hard to come up with things to do, dive deeper. Don't be afraid. What did you always dream of doing when you were young? What puts a smile on your face? If you're lacking joy and excitement, make plans and get away for an afternoon or the weekend, preferably alone. Remember, spending time with yourself is the only way to truly connect and learn what your passions and purpose are. Go to a museum, a park, schedule a dance lesson, try out a personal trainer, test drive a new car, take a walk, rent a paddle board, write something, draw, grab a fold up chair and sit by the ocean, schedule a lesson to learn an instrument, or do something you've never done before. If the idea of doing something new scares you, that's a good thing! Finding your purpose is about challenging your normal ways of thinking. Ask yourself what scares you? Is it being alone that you fear? Is it the fear of failure that holds you back? If it's failing to achieve a goal, that's normal, okay, and only proves how much it means to you. In Chapter 4, I will touch more on how to handle fear, but for now, all I want you to know is that fear is a time waster. Be liberated! Unleash your passions, take today by the hand, and lead yourself to where you really want to go.

YOUR SUPERPOWER

What I got and what I lacked in childhood turned out to be the same thing—I felt alone. I realized very early that I wasn't like Pop or my sister. I always fell asleep too early and took too much to heart. I didn't like roller coasters or scary movies like they did. I was different, not as tough. My mom was much softer than Pop, we were more alike, but she wasn't around. She was off living her new life without us. I saw her on the weekends, but in survival mode, I had shut her out and distanced her from my heart. I felt numb toward her, and trust did not exist between us. Unfortunately, pushing her away only added to feeling more alone.

Sometimes what you're missing can give you superpowers to do the extraordinary. I can't remember a time that I didn't feel alone. Because I lacked outside connection, I created one on the inside with myself. I turned my loneliness into my strength by spending hours getting to know myself on another level. I became my own best friend. It's because of this that I've always know with confidence, who I am and what I want. I know what my passions and purpose are. And the best part of all is, I know today that I am limitless. You can know the same thing. The first step is acknowledging the things you received in childhood and the things you wish you received.

What was the best thing you receive in childhood? What was something you wished you'd received?

_____

_____

_____

_____

_____

_____

_____

_____

_____

_____

_____

_____

_____

_____

_____

_____

_____

_____

_____

_____

_____

_____

No matter what you wrote above, you can expand your strengths by turning those very things into your superpower to help you on your journey. Going FSIR, whatever you lacked, realize that you found a way to fulfill that need yourself. That takes strength, courage, and intellect. Use this fact to create pride in yourself and build more self-worth.

What untapped potential is lying dormant inside of you? This is your journey to find your worth and your purpose, something that many people never do. Changing the way you feel about yourself will unleash what you really want from life, expand your horizons, and help you find true happiness. After you begin to love and value yourself, your purpose will become clearer and you will feel confident to go toward a life that makes you happy. Can you picture it?

Imagine you have all the money you need and all the education it takes to be whatever you wish to be. What would you do if you could choose any job, any place on earth, any hobby? Is there something you'd risk everything for? Is there something inside of you that you're not paying attention to? Have you ever wondered what a limitless you – looks like? For years I wanted to do more with my life. I've known for a long time that I had a very large calling inside of me. The older I became, the more I realized how much I wanted to become a motivational speaker. After losing my husband and realizing the FSIR method, I knew I was born to inspire people and now I'm making it my goal. What about you?

If you were limitless, what would you choose?

_____

_____

_____

_____

_____

_____

_____

_____

_____

_____

_____

_____

_____

_____

_____

_____

_____

_____

_____

_____

_____

_____

_____

_____

I love what you wrote above. You have wonderful ideas, and it is now time to stop dreaming about them and do them! Act. What you want is attainable, even if it scares you today. Especially if it scares you! Reaching for your dreams is how you will find your purpose, but there's one person who could stand in the way of your success; and it's *you*. To make your journey a success, let's make sure you're not stuck in something I call, autopilot.

CHAPTER 3

AUTOPILOT

When your self-worth is low, many people go into autopilot. It's living life by going through the motions, day in and day out, and just like the movie *Groundhog Day*. Nothing changes. If you are on autopilot the sad fact is that you're barely putting effort into your life. Remember my favorite question? What are you looking forward to? Now it's your turn to think about what you are looking forward to. You should be able to think of at least two things in your life that seriously excite you. If not, you're likely stuck tightly in your comfort zone and allowing life to happen to you instead of the other way around.

The comfort zone is poorly named. The word comfort should be reserved for things like cookies, melted cheese, hugs, your boyfriend as the big spoon, lying out under the warm sun, flights to breakfast in the plane, vacations, and your eyes closed with nothing on your mind. Let's be honest, a comfort zone is nothing like those things. It's not made of comfort; it's made out of fear and sometimes just plain stubbornness. You stay stuck in that zone to avoid failure, disappointment, and embarrassment; but it also holds you back from growth, experiences, adventure, and excitement. You may have talked yourself into believing your comfort zone is a safe haven; but what you're really doing is resisting any and all opportunities. Make no mistake, there is no comfort in a comfort zone.

A comfort zone isn't comfortable at all, it's rigid—what I call an FU zone. When you are stuck, rigid, and never open to new things, you say to the people in your life that you are not willing to change. You may have even said, "This is who I am, deal with it." The FU zone is a place of loneliness. You will find that you don't have many friends, your fun levels are non-existent, and people probably walk on eggshells around you. I know some people in comfort zones who are real pills. They go out of their way to make life difficult for themselves and others. FU zones come from fear. You may have tried something a long time ago that you bombed on, and it hurt so badly that you decided never to reach for something again. Rigid people's goal is to protect themselves, but what they are really doing is making sure they never expand out of what they know.

Did you know that you can get into a U zone by going FSIR? U zones are the opposite of FU zones. This is where you step outside of yourself and admit there are things missing in your life that you're afraid to go for. To find if you are in an FU zone, ask yourself if you're doing everything your heart desires? Be honest about this one. Are you doing everything that your heart desires? If not, why? Does fear have anything to do with your unhappiness and unwillingness to get out of your comfort zone? The time is now to admit that a change is necessary. The time is now to go toward your purpose. By creating a relationship with yourself, you'll find more self-worth, you'll begin to love yourself, you can stop being afraid of change and learn to incorporate more opportunities to grow, and best of all; you'll discover your purpose.

Take a moment and think about your ideal life, one without limits. What is your dream job? Dream relationship? Dream home (location, house, who would live with you, etc.)? Be honest with yourself and write it down below.

_____

_____

_____

_____

_____

_____

_____

_____

_____

_____

_____

_____

_____

What if I told you, everything you wrote is attainable? Everything. No one can hold you back except you and it's up to you alone to get it done. Letting your guard down and giving in to those things you wrote that spark your heart is the way to answer your calling. Those things are waiting for you to answer them. The first step to realizing happiness is finding your self-worth. After worth comes a confidence that will make you unstoppable as you move toward your desires. With increased self-worth, you will carry yourself differently and need less from others. Most importantly, you'll fear things less. You will become stronger, smarter, and braver, refusing to settle for anything less than what you truly want and deserve.

I'm not going to lie; it's scary to set out on a new journey. And I'm suggesting you do it alone. It's something you've probably never even thought of! Nurturing your self-worth will create a tremendous amount of satisfaction in a fresh, new you that you never even knew existed. To find this enhanced version of you, you'll have to get uncomfortable. You talked yourself into a comfort zone, and now you're going to have to talk yourself out of it. Getting comfortable being uncomfortable is just like getting into a comfort zone. You worked hard in the beginning when you set up your comfort zone. Now, you'll work hard on breaking it down and getting out. The amazing part of being uncomfortable is growth. When you're uncomfortable, that is when the most growth happens. To be honest, growth is what life is about. Remember to be kind to yourself. Be gentle with the new you. You've never been this far before. You are in unchartered territory and things are going to be different. When you are uncomfortable, try to sit with the feeling and ask yourself where it's coming from. Ask yourself where it comes from. Does your discomfort come from when you were younger? Do you think it comes from someone else in your life? Is there someone you don't trust? What parts of your life are you unsatisfied about? Are you disappointed in yourself? Are there things you wished you would have done by now? Do you blame someone else for where you are not in your life? Remember it's okay to feel discomfort. Try to see where it comes from and be gentle as you listen to yourself. Discomfort isn't comfortable, that's why you're in your comfort zone. The U zone is for you to gently break out and take control of the things you want for your life.

CHAPTER 4

YOUR BRAIN ON
CONFIDENCE

O k, so now that you're on your way to increasing your
self-worth and going confidently toward your purpose,
suddenly, you freeze. Your thoughts turn from limitlessness
to limited. You may think limitations only come from
not having enough money or time, but the truth is, most
limitations come from your own brain. Your brain is the only
thing that can make a change but it can also be your biggest
challenge to making changes! You may think you want to
make a change and then, suddenly, it will tell you all of the
reasons your change won't work! The reason is because you've
trained your brain to do what you want it to do. Its main job
is to keep you safe. It's main job for years may have been in a
comfort zone, and now that you want out; it doesn't want to
let you. You know that staying in your comfort zone is what's
limiting you, so, how do you overcome this contradiction?
You need to put your brain on confidence.

Since the day you were born, your brain has stored every bit of information it's ever seen, heard, tasted, smelled, or touched. Everything from your past rules how your brain reacts. Everything you've learned is stored inside your brain to help you survive. Your past dictates your future until you teach it something new. When you encounter a problem in your life, your brain immediately thinks back to a time it felt similar and reminds you, good or bad; what happened. For example, say you want to plan a trip somewhere you've never been before. Your brain will think back to the last trip you took. Everything that happened on the last trip will affect the choices you make for the next trip. Your brain is your guide, but it can only remind you of what it's already been through. It only has what it knows and it's very protective. Its job is to keep you safe, but it often overestimates threats without considering the potential rewards. The only way to make your brain react differently is to give it new information.

Have you ever noticed that some decisions are easier than others? Those decisions are so easy because you're confident about your ability to achieve your goal. With confidence, your brain will do anything you ask of it. The other decisions that you're not so confident about are harder to pursue. When you decide to go to college, change jobs, or learn something new, you are forcing your brain out of the comfort zone of what it already knows. You may have to grab a chair and sit down with your own brain for a conversation, but it won't be easy.

Let's say you want to return to college; it might go something like this:

You: Ok, listen. I want to get my bachelor's degree finished. I've always wanted to do this.

Your Brain: Um, no. That's not a good idea. It's too expensive.

You: I'm going to apply for financial aid.

Your brain: That's not the only problem! You must pass a foreign language to get your bachelor's degree.

You: Yes, but—

Your Brain: Do you remember how well you did in high school Spanish? You dropped out. How do you think you'll be able to pass a college-level foreign language when you couldn't even pass a high school class?

You: I will study extra hard this time.

Your Brain: It's going to take forever to get a degree! You don't have the time. And think about how far away the campus is? Fifty miles! That's a hundred miles per day, round trip. Besides the time that drive would take, gas is expensive. Not to mention all the mileage you'll put on your vehicle…

If you allow it, your brain will go on and on with everything it possibly can to steer you away from this decision. Even though your brain belongs to you, it can feel like it operates independently. Without confidence and new information, your brain will override you fast, and then your *full speed ahead* to being stuck again. When lacking confidence about something you want, ask yourself why. It's usually because of how you feel about yourself. Doubt comes from low self-worth. Everything you want is on the other side of feeling worth it; you must convince your brain.

Here's how to "wow" your brain with something new—your attitude is law. You constantly talk yourself into and out of things based on how confident you are. If you're not contemplating something new right now, it's because you're not expanding your brain with new information to make a change. Your brain remains satisfied because you've trained it to feel that way. If you don't fight against it, it will always take the easy road. Changing anything takes a lot of things: self-worth, confidence, and positive self-talk, just to name a few. Your thoughts must be infused with your own strong belief that you *can* accomplish those things. Growing your confidence comes from knowing who you are and believing you are capable and most importantly – worth it.

I know this firsthand. That example conversation with your brain from earlier? Yeah, that was a real conversation from my life. I told myself that I wanted a four-year college degree, and my brain talked me out of it. For years, I allowed my negative self-talk to win. Your thoughts are the only thing that will hold you back. You must train yourself to have positive inner dialogue.

Take a moment and think about the things you want. Listen very carefully to what you *tell* yourself about them. Self-talk is critical, so pay attention to what your inner dialogue tells you. Then remind yourself how badly you want it. Tell yourself you'll do anything it takes to get it. Say, "This is going to work." Your thoughts have everything to do with your outcomes because self-worth stems from positive inner dialogue. Every accomplishment you've ever had comes from how confidently you've talked yourself into it.

If you're stuck in a negative self-talk pattern, you will have to consciously shift your mindset. Instead of talking yourself *out* of the things you want, change how you talk to yourself and allow yourself to talk yourself into pursuing your dreams. Go FSIR and remember all the things you're proud of. Remind yourself how strong you are and how many difficult things you've made it through already.

If you're struggling to come up with things from your past that make you proud of yourself, another great tool is imagining how you will feel after one of your goals is accomplished. Using my personal example of the four-year degree, I imagined the feeling of satisfaction when all my classes were picked out. I imagined how proud of myself I'd be when I passed all of them at the end of the first semester. Then I thought about how it would feel when I was in my last semester and the satisfaction I'd feel when I was finished. I imagined the day my diploma would be in my hand as one of my favorite days. Once it's done, it could never be taken away from me! Because of my time and effort into changing my thoughts, I was able to talk myself into getting my degree by focusing on all the things I would have because of it. I reminded myself that I would be able to get a better job, make more money, and finally stop wanting my degree. And guess what? I did it! I got my four-year degree!

It's your turn. Remember your ideal life that you wrote down in Chapter 3. Now I want you to envision it coming to fruition. How will you feel *when* you achieve it?

_____

_____

_____

_____

_____

_____

_____

_____

_____

_____

_____

_____

_____

_____

_____

_____

_____

_____

_____

POWERFUL THOUGHTS

Have you ever noticed how some people attract things? The ultra-positive people's lives seem to brim over with positivity. They talk positively, think positively, live positively, and attract positivity. Then there are the Negative Nellys. Those people say things like, "I have the worst luck," "Nothing good ever happens to me," and "If it weren't for bad luck, I'd have none at all."

Your thoughts are *everything*. One thing I know for sure is that you get what you focus on. All you need to do is observe your thoughts. Try it for a day. What consumes your mind will fill up your day. I find the more something is in my mind, the more I see it in my life. A great example is a new car. Have you even thought about getting a new car? You sit at your computer and look at the new car on the manufacturer's website. I even go through the options and build the car I want, just to see how much it would cost if I found one with everything I wanted. The magic happens the next day, when I'm driving, and I see that car! Five minutes later, there's another one. I've even heard myself say, "I see them everywhere now!" Do you think that car just magically started to appear around me? No. That car was always there, I just didn't have it in the front of my mind. The same is true for all thoughts. Now, think of your job or your relationship. Those are two of the most important parts of many people's lives. If you are sitting at your job wanting to be somewhere else, or living in a relationship with someone you aren't happy with, I'd bet your thoughts are about all the negatives you're facing. Instead, try for a week to turn your thoughts around and focus on the positive; or go a step further and put out there what you'd rather have in your life. In no time, the things you want to see will appear.

By focusing your attention on the positive, you will see more positive things. If you focus on the negative, you will attract negative things. If you find yourself getting a lot of what you don't want, it's because you're thinking of those things. Even if you argue with me and say, "I wouldn't bring negative things into my life!?" You may not realize that your thoughts are exactly the reason your life is not the way you want it to be. Try to imagine for a moment that there's a chance you're addicted to those things in your life. Even if they are negative. Go FSIR and think back to a time when you may have said something like, "I'll be stuck at this job forever." or, "All men are the same, I'll never find one that is good for me." Even though you'd hate to admit it, you might be stuck at your job because you choose to keep going to a job you aren't happy with; you also might be great at choosing the wrong guy because you're living up to your own expectations... This is a giant concept. It might make immediate sense, or it might take a while to admit to.

Train your brain to think positively and you will get back positivity. I strongly urge you to try it. As you move forward, remember this exercise, and manually/continually refocus on the positive. Your brain will change through repetition. Speak positively to yourself every day, and soon, the positivity will show up, just like that new car. Take a look at what's going on in your life and realize you have more to do with it than you realize. Make it a habit of shining a light on your thoughts and changing them to what you want. This way of thinking will not only bring more awareness, but it will also spread to your goals and dreams. Before you know it, anything you want will become a reality. It's as easy as teaching yourself how to use a spoon... Yes. A spoon. ●

SPOONS

Have you ever met a functioning adult who couldn't use a spoon? Every adult I know is skilled with a spoon. I'd bet you're pretty good at using one too. At some point in your life, you mastered using a spoon. Today, you don't have to think about it—you just do it. A no-brainer. It's simple now, but it wasn't always.

With the passing of time, it's easy to forget where you started from and begin to take for granted how easy it is to perform a learned skill. The first time you held a spoon, you were terrible at it. You tried eating applesauce, yogurt, or mashed carrots as your mom or dad cheered you on. You could hardly hold the darn thing, let alone get the food to your mouth! Your fruits and veggies wound up all over your high-chair and little baby bib. It probably took you months to successfully learn all the facets of spoon usage—to hold it properly, to scoop with the correct side, and to get every drop into your mouth rather than on the floor. Yet, each day you practiced, you got better and better at it, and now, you're a master at using a spoon. This is true with *anything* you want to learn. You'll start off lacking, but the more you do and practice, the easier it becomes.

The process of learning creates a rut in your brain. That rut is called a neural pathway, and it's an actual path. Think of it like a four-wheeler path in the woods. If you drove over the same path every day, it would become well-worn with your tire marks. If you take this same path for a few months, you will create ruts. Imagine one day you try to drive out of the deep ruts, but they are so worn in by now that they are pulling you back onto the familiar path.

Your brain is no different, and this is how we get stuck in a rut—literally. Your repeated patterns and behaviors create ruts in your brain that make pathways. Each one of your behaviors, beliefs, and thoughts become neural pathways. The way you act is a path. The way you talk to yourself is another path. The more you do it, the easier it becomes. If you act in a certain way or believe something for many years, the ruts can become so deep those things seemingly happen automatically. They happen so often, they become second nature, just like using a spoon.

But let me clarify a major factor here: each action you take or word you speak is a choice. It may be difficult to pull yourself out of a rut, but it's not impossible. Use this knowledge to become great at whatever you choose to spend your time on and choose wisely. You can become better at anything, if like learning to use a spoon; you don't give up.

Is there something you wished you were better at? A particular hobby you want to take up, a sport, musical instrument, artistic style, a new career, or something else you've been wanting to try? Take your time and really think about this. Write down as many things as you can—I just hope it's not using a spoon!

_____

_____

_____

_____

_____

_____

_____

_____

_____

_____

_____

_____

_____

_____

_____

_____

_____

_____

_____

_____

_____

Anything you wrote above is possible. No matter what it is, you can learn if you dedicate time to it. Someday, you'll become an expert at it, just like using a spoon. If you disagree with me, remember that's your brain limiting you because you haven't taught it a new way to think…*yet*. It could also be the lack of self-worth that keeps you from believing in yourself. Fear is another rut your brain creates to keep your life predictable—remember the comfort zone? These ruts will keep pulling you in and telling you that you're not worth it, it will take too much time, you're not smart enough, and you don't deserve it. None of that is even close to the truth.

Go FSIR and get out of that rut! Imagine yourself not being afraid to try new things. You can learn anything you want to, and if you ever find yourself doubting this; just remember it takes time (like the spoon). If you start learning something new today, imagine how much progress you can make in a year. Build up your confidence and you will become unstoppable and limitless. You have the power to change any negative rut you've ever made in your mind and turn it into something positive. Becoming limitless begins when you let go of what you fear and begin learning something new.

OVERCOMING FEAR

Fear is a form of protection meant to keep you safe. It's there to caution you so you don't die, which is good in *some* circumstances. A thousand years ago, fear was valid when humans were hunted by animals. If you've ever seen the movie

The Croods, you'll remember the dad feared everything. He and his family lived in the stone age among predators, and he nurtured fear to keep him and his family safe.

Today, most fear is unnecessary because life is much easier in that sense. You don't have to kill your food and other things don't hunt you to survive. If you're hungry, you go food shopping or to a restaurant. So, even though fear is designed to keep you safe, it's when fear keeps you *too* safe that it gets in the way of progress and growth. When you fear too much, you hold back on every aspect of your life. Fears set up boundaries where purposes get lost, confidence dwindles, and finding self-worth and self-love feels impossible.

The only reason you can use a spoon today is because you never gave up on it. A key factor to never giving up is pushing aside your fear of failing. When you first learned to use your spoon, you were too little to be embarrassed and worry about failure. Unfortunately, fear of failing gets worse the older you get and the more self-aware you become. When you go to school and fail at something, your peers laugh at you. You learn what embarrassment feels like. Having witnesses watch you fail has a huge impact on what you're willing to risk next time. Even watching others try and fail can cause self-doubt to stir within. Failing can dramatically change how you see yourself.

If you lack self-worth, your fears can hold you back considerably. When you fail, it feels terrible, and your brain remembers that. That feeling will create a negative rut and it will stop you from going toward your dreams the next time because your brain will remind you how horrible you felt. Fear is the main reason you create a comfort zone. It stops your courage, clouds your purpose, and causes your brain to decide that a limitless version of yourself is no longer possible. Fear is usually the biggest hurdle that stops you from becoming who you want to become. It tells you that you can't have what you want, but that's not true. You just haven't taught your brain to push past that rut...*yet.*

Here's something that might blow your mind: did you know that every fear you have is created by you? Yep, all fears are man-made. Your brain creates fear based on your past to keep you comfortable and predictable, and even though you don't realize it, they keep you limited.

To erase fears, you need to talk to yourself differently. Your inner dialogue must transform from negative to positive. You need to encourage yourself to go for things you want and remind yourself that you'll never achieve your dreams if you always play it safe. How you talk to yourself is directly related to how you feel about yourself. Your comfort zone may seem like a safe place, but the only thing safe about it is that you'll remain exactly where you are. Become your own best friend. Tell yourself that you're worth it. Remind yourself how important you are and that you deserve to be anything you want to be.

How many minutes, hours, days, and years have you wasted fearing all the things that could go wrong? It's easier to talk yourself out of something than it is to talk yourself into it, but it's worth the effort to create a new pathway and discover the opportunities that await. *You* are worth it.

Going FSIR, take all of what you learned so far to erase your fears. Here's how to stop fear:

1. Identify who you are.
2. Find and create self-worth through self-pride and thrive.
3. Let go of your past. Only use it to help enhance your present.
4. Spend quality time alone and search within.
5. Get off autopilot by finding your passions.
6. Discover your purpose and pursue it.
7. Develop positive inner dialogue to encourage yourself.
8. Grab a spoon and try something new.
9. When fear tries to talk you out of something, overcome your brain with confidence.
10. Never give up.

The Buddhists believe that fear of the future is pointless because the future hasn't happened yet, so it doesn't exist. They believe fearing your past is pointless too, because the past has already happened, and it *no longer* exists. I love this concept! The past is done and gone, and the future is what you make it. Fear does nothing for you. Simplifying fear to this degree could potentially make everything you've ever gone through seem obsolete.

Do not think for a second that I'm downplaying what you've been through or how terrifying fear can be. Some of you have been through traumatic events, and that type of fear is far from simple. I've spent hours in counseling working through my own trauma. Often, people will suggest simply letting it go. That solution is frustrating to hear. If it was that simple, I would have done it a long time ago. I'm sure you would have too. Some problems cannot be solved that way.

Trauma creates a wound in how your brain processes things. Some of those wounds require professional counseling to heal, but healing properly will allow you to let go of the hold trauma has on you and learn how to process it differently. It's also important to develop a strong support system. When you have people you trust in your life, it can be a saving grace when fear overcomes you. It is very hard to ask others to help, but I urge you to do so. Asking for help is not a weakness. In fact, it takes a lot of strength to admit you need help.

Even if your fear is a more simplified version, it can still feel debilitating when it takes over your mind. Letting go doesn't always seem like an option. No matter what source your fear comes from, when it rears its ugly head, it may only last about ninety seconds. It cannot overtake you unless you let it. Your brain can only remember your past, so it's up to you to be courageous enough to show it a new possibility.

Deep wounds from our past take time to heal, and healing can only happen when you learn new information. Everything you're learning is a journey. Fear is also a journey. Go FSIR and begin to let go of your past by learning new ways to find self-worth. With new information, the future will become less cumbersome. Learning ways to process all your fears will release you from the hold it has on you and allow you to worry less about the future; all while being more present in the here and now. Living in the present is an art that makes life much more enjoyable, but like all art, it must be practiced for your skills to grow.

Every time a fear comes up, pause, and question it. Ask yourself where is this fear coming from? The discomfort you're feeling today may stem from an experience in your past. If your brain links today's obstacle to an old fear, it gives it more power. By going FSIR and identifying where your fear comes from, you can learn about it and how to overcome it. Use this time as an opportunity to be kind to yourself and seek understanding within. Remember you've overcome all the obstacles in your past. Give yourself credit for going through it, both then and now; and take time to learn what you need to let go.

It can be difficult to analyze fear when you're deep in the feeling. The next time you're amid feeling fear, try this:

Close your eyes and take a deep breath. Acknowledge your present fear and realize it cannot overtake you. Understand that old wounds can trigger and intensify new fears. Use this knowledge to remember that you've been through hard times before and you made it through. Find a connection to yourself by admiring your strength. Feel your worth. Feel how important you are.

Feel yourself remember the beginning of your fear. Seeing that it may come from a long time ago, it is only triggered today. Use this new knowledge to disarm your fear today. Old fears are over. Today is a new day. You have the power to make your own choices today. Look in your heart and listen to what it needs. Let it know you are listening and brave enough to think differently. New outcomes are possible with new, gentile information.

This exercise can do wonders in creating self-worth. If you allow yourself to take credit for surviving your past, you can take pride in yourself for having conquered difficulties before. You were the person who got up every morning and lived your life despite all the pain you felt. You can be proud that you're still living today. Your very life is a major accomplishment. Every breath you take is another second you have lived with the fears that have tried to overwhelm you. Another second will go by, and you will get through that, and then another. Honor yourself by continuing to move forward. You are worth it. I am so proud of you.

After learning what you need and being brave enough to go towards it, who are you going to be next? The choice is yours.

HEALING

I remember loving being alone when I was little. After my husband died, I was suddenly single and alone again. At this time in my life, I became afraid of being alone. I thought my fear was due to getting older, but it was because I felt insignificant. I hated to admit it, but I couldn't understand why my husband would replace me. I went to great lengths to make sure we spent time together because I could feel us disconnecting. Since his death, the feeling of insignificance has heightened. The root of the discomfort comes from when I was little, and felt very unimportant to Pop, my mom, and my sister. They all had so many other things to focus their attention on. I do remember filling up my own needs as a child, but Shannon's death has left me with the feelings of insignificance I have felt in my life. It expanded them. Going FSIR is painful sometimes, especially reminiscing back on feelings of lack and then adding grief to those feelings. When my mom died, I could only remember her good and my bad. I went to see Shannon's body to help me remember that he pushed his life that far. I know healing is a choice. I know now that I must allow all the feelings in, good and bad. I realize in focusing on the bad, it leaves me wishing for things I cannot change. In remembering the good, it gives me satisfaction that I did my best.

After Shannon died, to avoid being alone, I chose to let in the wrong people in my subsequent relationships, and in each of them; the feeling of insignificance resurfaced. My usual reaction was to run because I was scared of being hurt again. If I ran, the choice would be mine to end things, and that would save me from getting hurt. But shortly after making this decision each time, I would realize I was single and alone again. And the cycle would repeat. Going FSIR, a life coach I was working with asked me to dive deeply into remembering the first time I felt insignificant. Easy—it was when my mom moved out when I was four. That feeling intensified years later when my husband chose another woman and then died, leaving me completely alone to raise our sons. The parts inside of me remembered how badly it had hurt to feel so insignificant to two very important people in my life. I felt threatened in my current relationships whenever that feeling of insignificance arose again—a feeling that had the memory of my mom leaving and my husband dying to strengthen it. Through this exercise, I learned how to stay with the fear whenever it arose to see where it really came from. My desire to run stemmed from my unhealed wounds that had created deep pathways in my brain.

I went FSIR again and realized my fear prompted a negative feeling, but I didn't have the tools I needed to fix it. I learned to ask for help and turned to counseling. Ever since, I have been in the process of healing those parts through counseling, deep understanding, vulnerability, and increasing my self-worth. I have been learning new solutions to heal the feeling of insignificance. It has been scary to touch things I don't understand, but it's also been liberating to understand how my mind works. I have new tools and the support of my loved ones to change the negative behaviors that held me back. At last, I am healing from wounds that are decades old.

Sitting with difficult feelings isn't pleasant, but it is a necessary part of healing. When I first sat with mine, I was worried that I wouldn't find the answer. I closed my eyes and listened to the fear inside. It hurt. I discovered that it came from my past. But I'm not four years old anymore, and it's been almost ten years since my husband died. I've come so far since both of those events, but events from my present trigger them. I've gotten through some very difficult feelings. Counseling provided me with new tools to help my wounds heal. I've made new neural pathways in my brain that are becoming easier with repetition. You can do the same by gently sitting with yourself and listening to your old fears. It will allow you to disarm them.

You might not like the idea of counseling. I didn't want to need it either. I was afraid to need it, as though seeking counseling would be admitting that I was damaged or broken. But I decided to take my own advice and remember that my fear was created by me. I also knew the wounds in my brain would continue to repeat if I didn't get help. Counseling has such a stigma. I wish it was called healing instead—then it would feel special and get the recognition it deserves. Asking for help when you need it makes a difference as long as you do the necessary work.

When I went deeper with a counselor who specialized in trauma. I learned about internal family systems and about the parts inside of me. People who haven't experienced trauma have internal parts that work together well, while people who have been through trauma have parts that become conflicted. The main goal is to listen to the wounded pieces and give them confidence that who you are today is okay. In fact, who you are is strong; you are someone who has survived trauma.

Who you are tomorrow could be someone even greater, stronger, and braver—someone who has not just survived trauma but someone who is conquering it.

I have been able to listen to the parts that have been hurting inside and finally find healing. This process has been magical. I encourage you to seek counseling and start the eye-opening healing process for yourself. When feelings of discomfort begin to overtake you, pause for a few moments, and listen to yourself to see where it comes from. Be patient with yourself. Stay with it for only 5-10 minutes and write down your thoughts. What triggers you? Go FSIR and ask if you've felt this way before and if so, when? If you've felt this way before, are the current choices in your life the reason it keeps happening again and again? For example, are you addicted to something? Like the belief that all relationships fail, but not realizing that you keep choosing the wrong person, and when things don't work out, it proves that you're right? Or does your discomfort come from a long time ago and you're scared today because you don't want your past to repeat itself? The past can be tricky. Try to split today's discomfort off from your past and just deal with today as a separate issue. Or you can create a positive trigger...

TRIGGERS

I developed a new trigger this year. I didn't see it coming since I have worked hard to become confident and secure, so it was the last thing I expected. Since meeting my fiancé, something inside me changed. He and I have different stories. He shares custody of his two teenage children with his ex. I don't have another parent who's with my sons when I'm gone. On top of that, my fiancé travels hundreds of miles each week for work. Although he is home most nights, we live over an hour apart, so our time together is limited. To make up for lost time, we Facetime each other every night and I share all the details about my job, my sons, my friends, and everything else I can think of to tell him about my day. He listens, but he volunteers very little information about his own day. He keeps his feelings inside, and this has become a trigger for me. When I don't have input, my mind goes to work. I wonder who he sees every day. He's been a traveling sales rep for almost thirty years. He knows lifetimes of things about his customers. I know nothing. I have a new trigger that wonders if there's something he's hiding from me. It comes from when my husband who died, met a beautiful new woman, and started a relationship during our marriage. I am frightened to have to go through that pain again.

To use FSIR, there are three things you must do with a trigger:

1. Identify the trigger.
2. Ask yourself where the uncomfortable feeling comes from.
3. Find a way to control/manage your trigger.

In my case, my fiancé lives a life that is separate from me. I feel scared he may have things he's hiding from me. I have never been a jealous person; I would never be in a relationship with someone I didn't trust one hundred. However, after my husband replaced me with someone he worked with, I developed a trigger where I fear being blindsided.

To resolve this trigger, I told my fiancé how I felt. I explained that I'm not jealous; I'm terrified of being hurt again. He said he's not hiding anything and explained that he's simply not used to sharing. Now that he is aware of how it makes me feel, all I can do is hope he comes through. I told him that I'd love it if he shared stories about his customers and tells me more about his days. My hope is for my trigger to lessen because I am worth it to ask for what I need. I hope together we can find a solution.

Triggers are things that cause you to lose your balance and knock you off kilter. It's not easy to analyze things when you're actively amid a trigger, so try analyzing your triggers during a time when you are not directly affected by them. Any discomfort in your life is up to you to explore and change. If your trigger is too large to take on alone, ask for help. Talk to someone who shares the problem and see if you can work through it together. You can have control over your triggers, but you will need to offer yourself kindness and awareness; if you can't figure out how to control them, seek help. For years, I wasn't aware that I could control them. Now I've learned that being aware of them is the first step, it doesn't mean you have to figure them out the very first time; it simply means there will be things you can work on changing.

The Buddhists know how to handle this clearly. They believe pain will happen, but suffering is a choice. Fighting against something you can't change is like fighting the universe. You will suffer and you will lose. But you can choose to be aware of the things you cannot change, and in doing so, find peace. I love how simple that is and how it puts the responsibility back on you. Suffering is unnecessary. Make the choice to fight against it by putting your effort into understanding your trigger and finding awareness. For more on this topic, please consider reading the book, "The Untethered Soul," by Michael A. Singer. This is a life-changing book that teaches how to find awareness and stop the persistent cloudiness of your own thoughts.

I'd like to take you FSIR and teach you how to create a positive trigger. But first, you need to acknowledge that where you are today is the result of every decision you've ever made. You might want to blame someone else and say you're where you are because of someone else, but that isn't true. You know that as well as I do. Everything you do is because you thought about it, decided to do it or not do it, and now you're here based on that decision. Your decisions are driven by your emotions. So many decisions feel instantaneous, but they're not. Decisions and responses only *seem* automatic because they happen very quickly. Remember no-brainers? These occur when you are confident about the decision or action. The things you do repeatedly feel automatic because you do them over and over. It's the same with emotions. Realizing that every emotion you feel is a choice takes away the excuse that it's automatic and uncontrollable.

Now that you understand that let's create a positive trigger to override a negative trigger. Think about something that triggers you negatively. Focus on it. Close your eyes and really feel that trigger. Feel the heaviness in your heart and the frustration. Now, think of a time when you remember feeling happy. The day your husband proposed, the day your kids were born, the day you graduated from high school, or a time when you didn't think you could possibly be any happier. Think of the people you were with, how good it felt to smile, what the occasion was, how much joy you felt, and how happy your heart was. Feel how that happiness makes your heart lighter. Smile and feel the happiness spread all over you.

Physically choosing to turn your thoughts of a negative trigger into a positive one can yield incredible results. Be grateful for how fortunate you are to be becoming aware of your triggers and learning how to change your energy toward being happy. Even though you went through hard times, you survived, and if you choose to refocus your negativity into positivity, you can even thrive. Every emotion is a choice. You allow them to come and go. You give your emotions power by giving your focus to them. Nothing you feel just happens. It's your choice how you think, what you think about, and how you use your energy.

I used to be a mentor at a youth group. There was a young boy who drove me crazy! He was super dramatic and was always seeking attention. As awful as this sounds, I found myself wanting to avoid him, but as a mentor, I couldn't. He was there to be mentored. So, I decided to sit down with the boy and listen to him. He told me about his family. His mom and his dad had remarried other people and they each had more children with their new spouses. This boy felt forgotten at home because both of his parents were focusing on their new families. He was needy and struggling to get attention. He wanted to fit in and he was trying everything he could to figure out where that was.

Often, the reason other people set triggers off inside of you is undealt with pain that comes from your past. When you are uncomfortable or annoyed with someone, it's likely due to an insecurity inside of yourself. While I listened to this boy, I recognized something we had in common. In my childhood, I too felt alone and abandoned. After our talk, my heart changed. I softened toward him. I felt terrible that he was struggling and that all he had was a stranger to tell his worries to. I created a positive trigger for him out of the soft spot I'd created in my heart. This boy needed to feel special, and he found it in my positive trigger.

Breaking down your triggers creates an understanding of what others are going through. When all you focus on is your own discomfort, you will never be able to see the other side. Identify your triggers as something inside of your own heart and know they can be changed with an active choice to be aware of them. This was one of the coolest realizations I've ever had. Being aware of your triggers and softening to other people allows new information to get into your brain. It's the only way you'll ever be able to think differently about your triggers.

UNTRUTHS

My sister is funny, kind, and a very good person today, but when she was younger, she was a tough kid. When my sister was little, she believed she was to blame for my parents' divorce. It wasn't her fault. Our parents' relationship just didn't work out. After our mom left, it seemed like my sister was always in a bad mood. She developed a hatred for school and the older she got, the more difficult she became.

What she was lacking was positive feedback and positive attention. Every day, she heard my dad tell her that her attitude stunk, she didn't apply herself, and she was lazy because her room was dirty, and her grades were bad. She heard over and over for years that she was these negative things, so she believed them. My sister became a product of an untruth. It's when you become what you hear you are, not what you really are. She wasn't a bad kid; she simply lacked attention and connection, so she acted out. Without hearing positive things about herself, she believed the things my dad was frustrated about. If she had heard that she was good over and over, things would have been easier for her; and that's what she would have become. Both my dad and my sister were going through a rough time. They both needed each other and neither of them knew it.

If you have someone in your life that tells you negative things about yourself, I am so sorry. Untruths are hard *not* to believe, especially when they come from a parent or someone you love, trust, and look up to. If my sister had heard that she was smart, great in school, and that her parents were proud of her, she would have acted on those positive things. Sadly, that's not what she heard...

When I was in grammar school, there was a class called Gifted and Talented (G&T for short). There's a time where every kid believes they are gifted and talented, but when I wasn't chosen for that class, I formed the belief that I was not gifted or talented at all. I watched other kids in my class leave for their G&T class time and it set me apart from them. I felt average, no; I felt less than average. They were special and I was not. I believed those kids were smarter. I believed those kids could take harder classes because they had something I didn't have.

Even as I got older, I made myself believe that I wasn't cut out for college or college-prep courses. I wasted every one of my school years barely getting by, believing I was merely average, until my senior year. That year, for some reason I decided to sign up for two Advanced Placement college-prep classes, AP English, and Anatomy & Physiology. I did it to be in some classes with my best friend. I knew they would be harder than any class I'd taken so far and I was nervous, but I was also very excited.

In anatomy and physiology, I did great throughout the entire class. I loved everything we learned, and I was pretty good at it! The English class was different. I wasn't as interested, so I coasted through the first quarter. Surprisingly, I passed, but by the second quarter, I was failing. I wasn't putting in the effort. Then something happened that had never happened to me before—someone called me out on it. My English teacher pulled me aside and told me he was disappointed in me. He said I was wasting my talent. I was a senior in high school, and this was the first teacher to ever tell me I had wasted talent. He was right. I was wasting every bit of my potential coasting and believing I was average. That truth bomb hit me hard. He also told me I wouldn't be able to get my grade up enough to pass. That was a real eye-opener. After that, I dove into that class and learned everything I could, determined to prove him wrong. At the end of the year, I passed! I was on fire!

Sadly, it was my senior year. It was too late to be on fire. With my grades, I was headed toward the path that included an entry-level job and community college. There was nothing I could do to change the past. I couldn't go back and do it again. I had to accept that I had chosen to waste the last decade merely coasting when I could have done so much more. Going FSIR began by not getting chosen for Gifted and Talented in fifth grade. No, to be honest; that was just an excuse. The real reason was because I believed an untruth and allowed it to set me back for years. My senior year taught me that I was only as talented as the effort I put in. You get what you give. If I wasn't putting in extraordinary effort, I wouldn't get extraordinary results. It was that simple, and yet it was a hard lesson to learn. It was up to me to do more with my life from here on out.

An untruth is something you believe about yourself that isn't true. These are things we learn to become from anyone who has a strong presence in our lives, the things you hear over and over about yourself. *She's not very bright. He's unlucky. She isn't good at math. He's always messing things up.* The more you hear them, the more you believe them. As these untruths create a neural pathway in your brain, they become truths. We start to give in to what we're hearing, behaving more and more in the way people expect and creating a bigger pathway for that untruth. The bigger the pathway, the harder it is to stop believing it.

Here's an interesting way to prove this. Try to write with the hand you don't normally write with. Does your handwriting look like a child's just learning to write? That's because, in a way, your off hand is exactly like a child just learning to write. Your neural pathway is set up to use the hand you taught yourself to use. Writing with the opposite hand is difficult because you have never practiced it like you did with your dominant hand. Just like an untruth, if an action or thought is repetitive, the pathway gets stronger and trying to go in a different direction becomes more difficult.

With some effort and repetition, you can take the power away from your untruths and choose not to believe them by going FSIR. Remember spoons? If you began practicing with your off hand, you would eventually be able to write just as well with that one as with your dominant hand. And this is true for anything. If you practice long enough, choose to believe in yourself, and never give up, you will succeed in time. Instead of giving in to your untruths, train your mind to believe what *is* true—that you are limitless.

There are three things we can do with an untruth:

1. Believe it.
2. Learn something to change it.
3. Unlearn something to change it.

An untruth can only hold you back, if you let it. Some untruths require learning something to change it like believing you are bad at playing guitar can be changed by taking lessons and putting in the effort. Imagine if you started spending half an hour practicing and *learning* how to play guitar each night. How would it sound when you played the guitar a year from now? You must start somewhere. You will get better in time, but only if you put in the effort and believe in yourself. Go FSIR and decide you want to learn something new. Effort equals results. Want proof? I taught myself how to ski at forty-four years old. Three years later, I am now a very gifted and talented skier because I put maximum effort into learning it. The people who are exceptional at what they do have spent enormous amounts of time creating deep neural pathways to be good at those things. Strong desire paired with effort and practice will result in learning something new. Strengthen your neural pathways and you'll be flying down the slopes in no time. It's that simple.

Some untruths require unlearning, like negative body image, believing you're lazy, or that you're an angry person. Instead of putting effort into it, you want to remove focus so the pathway will get smaller. Take anger, for example. Some people get angry easily. If you take your mind off anger for a few minutes, you may realize that most anger stems from something else, usually something from our past. Remember all your emotions are choices. If you've been labeled as an angry person, then anger has a giant neural pathway in your brain. You'll need to unlearn this untruth.

The first day you try to unlearn something, it will be hard. Your pathway for this untruth is deep, so it may even feel impossible. Just like using your non-dominant hand to write with or playing guitar for the first time; it will be difficult, but it is not impossible. It'll take a bunch of small choices to divert your mind away from that well-traversed pathway, but you can do it, and soon, your new pathway will be bigger than your old one.

If you find changing your focus seems nearly impossible, try hanging out with it. Acknowledging it and spending time with your untruth will allow you to see where it comes from. For example, if you are quick to anger, sit still with this emotion whenever you're feeling it and ask yourself where it comes from. You might be surprised to learn that your anger isn't really anger. It could stem from frustration, disappointment, anxiety, or a negative feeling tied to a past event. Be honest about your feelings and give them your undivided attention. When you take time to acknowledge them, rather than feel terrible about them; that act takes the power away from them.

When trying to identify an untruth, ask yourself when the first time was you first felt this feeling. This will reveal where it stems from. When your anger surfaces again, acknowledge the fact that it may not be true; and it was from a long time ago. Be aware that something today is sparking it and bringing it back to the surface. The thing that makes you angry today might come from a long time ago – a time when you bought into an untruth. But remember how strong you are today and that you are building self-worth. You don't have to believe untruths any longer. Realizing that they aren't true may change the feeling entirely. If you keep working at it, your new neural pathways will soon be bigger than the old ones you wish to change.

Do you hear negative things about yourself from people in your life? What do you truly believe about yourself, good or bad? What untruths about yourself would you change if you could? Write them down and see if there are ways you can start to unravel these harmful untruths.

_____

_____

_____

_____

_____

_____

_____

_____

_____

_____

_____

_____

_____

_____

_____

_____

_____

It's possible to change untruths. It will take time, so don't be discouraged. Your brain is a tool and it's comfortable knowing what it knows. You will have to force yourself to stick with an uncomfortable feeling long enough that you fully understand it. Allow yourself to believe that it is possible to make changes and create new neural pathways. Stop giving power to your untruths. The sooner you do, the sooner you will see more self-worth and a new, much happier way forward. Remember, whatever you give your time to will always win.

Allow me to prove my point. Imagine something you've always wanted to be—a guitarist, an avid skier, an author. Now imagine that you *are* whatever it is you wish. Ready? Say it out loud. "I am a _____(insert what you've always wanted to be)." I have a feeling that, almost immediately, your brain shouted out in doubt. Now say something that's an untruth, like "I am lazy." I'd bet you had an easier time believing the untruth. Your brain is your biggest ally *and* your biggest enemy. You will be the first one to talk yourself out of the thing you want, but you can also talk yourself into anything you put your mind to. Your brain does exactly what you tell it to do. The key is to be consistent and never give up. The minute you give up, you'll be no closer to your dreams or finding happiness. But if you start working toward changing untruths, the further away from them you will be in a year and the more self-worth you'll have! You can become anything you want to be if you can just overcome your untruths.

No matter what anyone else has ever told you, I need you to know that you are an amazing person. You are here on this earth for a purpose no one else can fulfill. You are doing great things with your life, and you are going to do even more. You are important, smart, kind, and wonderful. I believe in you.

Please turn a corner down on this page, and whenever you don't feel worthy or need an extra boost, read this to yourself:

You are worthy enough to do more. Increase your value by improving how you feel about yourself. You can change your untruths by making new choices. You can be anything you want to be. You are strong enough to give yourself the time and effort to learn new things. You can accomplish anything you put your mind to. You are worth it. You are invaluable. You are limitless.

THE PROCESS OF BELIEVING

Since the day I graduated high school, I knew I wanted a college degree, but for a long time, I did nothing about it. Knowing what I wanted was only the beginning. The vital part is doing it. I waited twenty-six years before I finally pursued my degree.

When you set your mind to something, it means you feel worth it, and you believe you can accomplish it. Self-worth will allow you to intentionally develop yourself. You can't just *want* things to be different in your life. You must actively pursue them. It's the difference between mindlessly scrolling through Facebook or using that time to advance your life—take an online class, complete that project you've been putting off, or apply to a new job. Once you know what you want, the next ingredient is feeling worth it, but the most important factor of all is taking action to get there. All are required for growth.

Remember, this is the step *before* self-help. It's about building self-worth first. You're never going to change any of your thoughts or actions without first believing you are worth it. This book is about finding your *why*. Why should you change anything? Because you're worth it! If you're unhappy, it's because you probably don't feel worth it and you're not confident enough to go for your dreams. When you go toward your dreams without restrictions, that is where you'll find your why - your purpose. Doing the same things you've always done won't change a thing. You must feel your worth, discover what you want, and put in the effort to achieve that dream. When *you* become the reason you change, everything about your life changes, and that is where the magic happens.

Step 1: What do you want in life that you don't currently have? (Career, relationship, skill, living situation, etc.)

_____

_____

_____

_____

_____

_____

_____

_____

_____

_____

_____

_____

_____

Step 2: Write down all the steps it will take to get exactly what you want.

_____

_____

_____

_____

_____

_____

_____

_____

_____

_____

_____

_____

_____

_____

_____

_____

_____

_____

_____

_____

_____

_____

Now, here's the hard part: look at your plan and begin. Do you know the one difference between the people you admire and yourself? Effort. Those people you admire are so worth it to themselves that they put tons of effort into what they do. Their worth allows them to go toward their purpose. You know who they are—pro athletes, musicians, dancers, actors, doctors, specialists, motivational speakers, and the like. They're the ones living with purpose and passion. They knew what they wanted, they felt worth it and believed they could achieve their goals, and they actively pursued them.

Even ordinary people can put in extraordinary effort. Stop holding yourself back and take that first step today. Believe in your worth and it will become easier to go toward your purpose. Do not waste time telling yourself all the reasons you shouldn't pursue your dream, because the sooner you begin, the sooner you will realize your happiness. Yes, it may seem daunting, but you are learning new information that will allow your brain to override its old habit of reminding you of all the reasons to believe your dreams won't work. Simply begin and move closer to your goal every day—even a little progress, is still progress. It will become easier and easier to go toward the life you desire, and soon, you'll wonder why you waited so long.

Believing you can do it comes with your increased self-worth. Do not stop until you value yourself and then you will move confidently in the directions that make yourself proud. This book is living proof. I did not stop until I got published. I quit my job to write this book because I believed that it would be valuable to you. I put massive effort into this book. I wrote even when I didn't want to. I pushed past the desire to fold laundry when my brain made it sound more appealing than writing. All because I believed in my worth. I believed in myself beyond just wanting something—I actively pursued it and I did not give up.

It's your turn to push past being tired, uninspired, and listening to the voice that says you're not good enough. Push past your brain when it tells you that someone else is smarter than you, has more talent than you, is more educated than you, or has more passion than you. Tell yourself that you are worth it. You have a unique value that no one else has. Your passion could very well help dozens, hundreds, or even thousands of others. Know what you want and do something about it.

Are you unhappy and unfulfilled? Take a moment to answer these questions honestly.

- Do you wake up each morning excited to start the day?
- Are you giving 100 percent to your job?
- Are you in the relationship you've always dreamed of?
- Are you satisfied with your life?
- Are you living your purpose with passion?
- Are you the person you were born to be?
- Are you truly happy?

If you answered no to any of those questions, there's a good chance you're settling for less than you desire. You're most likely settling for less than you deserve. Settling will never get you where you want to go. You are missing your best life. Think about what you could do today to get closer to what you want. Think of what you could do today to make tomorrow better. Are you waiting for the time to be right? If you're unhappy with any part of your life, the right time is *right now*. If you'd started five years ago, you might already have achieved your dream life. But we can't change the past, so all you can do is start today and get that clock ticking. By starting today, envision how far you'll go in a year! Believe that you are worth everything you want out of your life today and do everything it takes to get there. It is never too late to become the greatest version of yourself.

THE POWER OF A POSITIVE MIND

Have you ever thought about someone and minutes later they call you? That is not a coincidence. Your thoughts are that powerful. Thoughts are magnets. They send out vibrations and connect you to other things in the universe. Your thoughts can either create magic or havoc. The choice is yours.

While all my high school friends were heading off to new towns to start their exciting lives in the dorms of four-year colleges, I was stuck mucking stalls and working at a bakery in Bordentown, New Jersey. To top it all off, I was unhappily single, solitary, and feeling left behind. I was feeling worthless and very sorry for myself and looking for my worth on the outside. I wanted so much more from my life. Feeling everything converge upon me that day, I dipped into the bathroom and began to release my sadness. Looking in the mirror, I said, "My life couldn't get any worse." About three seconds later, I heard a loud crash from outside. I instantly knew what happened. My truck had gotten hit while parked on the street. I brought that thought into the world, and immediately, something worse happened. Your thoughts are the most powerful things on this earth. Need more proof: this past winter, my son was at our neighbor's house during an ice storm. They talked about the possibility of him crashing in the driveway on the ice. He didn't think much of it, but while he was pulling out, he ran off the driveway and crashed. They thought it, and it happened. What you think—good or bad—comes true. Make sure you're thinking good thoughts.

I quit my job to write this book. After eleven months, I needed to get another job to start making money. I was focusing my energy on not having a job, and every day, I received rejections on my applications. I was getting exactly what I was thinking about – not having a job. Until one day, I went FSIR and closed my eyes. I dreamed of my perfect job and how good it would feel to make money and work somewhere I wanted to work. In less than two weeks, I had a first, second, and third interview that resulted in an offer for a wonderful job.

Positive people are surrounded by positivity. They can't help it. They feel lucky and they are lucky. They are always brimming over with abundance. They smile all the time. They lighten the moods of those around them and draw others in. On the other hand, negative people are always in a bad mood. They think things can't get any worse. They believe they have bad luck, and it seems as though no one wants to be around them. They create sickness with a negative mind and stay stuck in poverty by thinking they are poor.

The universe does not care if your thoughts are positive or negative, you will get back whatever you put out into the world. Turn your thoughts around and create perfect health with a positive mind. Create abundance by feeling you have more than enough money, love, and happiness. Your thoughts work like magnets, so be very careful with how you use them.

Have you ever wished for something? Wayne Dyer teaches to "think from the end." Wishing is just a wish until you can imagine yourself *having* what you wish for. When you can imagine the feeling, you will feel when your wish comes true, the result is receiving your wish. At the end of the day when you lie down to sleep, replay the day, imagining each part of your day as though it happened the way you wished it would. Feel how you would feel if your life was the one you dreamed of. Do this before you drift off to sleep each night, and in the days that follow, you will find more and more of what you wish to be coming true. You will be amazed at the power your positive mind can have.

I used to worry internally and whenever I do, I get shingles on my lip. I have learned to focus on the power of my positive thoughts to keep myself healthy. Now when I am stressed and I can feel the tingle of the shingles beginning,

I immediately turn my mind to positivity and focus on the absence of shingles. I say to myself, "I am shingle-free." I think of my skin and see it healthy and perfect. Most of the time, by the end of the day, I can make the tingles go away. When I find it particularly difficult to stop feeling stressed, I can at least make the shingles a lot less intense by focusing on perfect skin and perfect health. I can also do this for my sinus problems. I might be allergic to cats, but I love our outdoor kitties. When I feel myself start to get stuffy and sneezy, I tell myself that my sinuses are clear. I am in perfect health. Usually within thirty minutes, I am clear.

I have even sold my house using this method. Two years after Shannon died, I needed to move out of our big house. The bills were very high, and I didn't want to be married to the mortgage and taxes. Besides, this big house was never my dream—it was Shannon's, and he was no longer here. It was time to let it go. I put a feeling in my heart that there would be three couples who would come to see my home. I knew the first two were not going to be the ones. Like magic, the third couple who came to see our home bought it for my asking price. I also imagined that it would feel like I'd known them forever—and they would be the perfect buyers. I didn't want to use a realtor, so I listed and showed the house on my own. It was amazing when they showed up. I truly felt like I had known them my whole life *and* they bought it for the asking price. They not only let us stay in the big house until our little house was finished, but they are also still wonderful friends!

Use your powerful mind to envision what you want, focus on it, and believe what you want is possible. Because it is.

CHAPTER 5

SUCCESS & FAILURE

I graduated from high school in the middle of my class in 1993. I loved school but never gave it my all when I was younger. Because I believed I was merely average, I showed up and only did the minimum. Secretly, I dreamed of living in a dorm at a four-year college and waking up to walk to my classes. Back then, that would have been the definition of success for me, but I didn't do anything to get there. I wanted it, but I didn't really work for it.

All my mediocre work got me middle-of-the-road grades and a ticket to community college. This reinforced my belief that I was average. I saw myself as a failure. I had thrown high school away, wasting years just coasting along without knowing my purpose. After graduation, I headed to community college and did the same thing I did in high school—I coasted. I didn't feel proud of myself. I studied only when I had to and completed sixty-five credits. My life was turning out average...again.

85

And then I gave up on school all together. I worked more hours and decided to not enroll for the next semester. Time ticked by. I met my husband and got married in 1999 at the age of twenty-three. Six years later, we had our first son, and two years after that, we had another. I continued to dream of going back to school and getting my bachelor's degree, but every time I contemplated it, I talked myself right back out of it. I reminded myself of all the reasons it wouldn't work. I told myself it would be too expensive, I probably couldn't pass a college-level foreign language; and since I had two kids, it would be too hard.

But I still wanted it. When I turned forty-three, I finally realized time would pass whether I went back to school or not. I would continue to get older; the only difference would be whether I got a degree. I went FSIR and decided to figure out how to go back to college. I wanted a better paying job, and I needed a degree to achieve that goal. I believed I could pass any class, and although I was worried about Spanish, I decided I had to do it. I wanted my degree more than I was afraid of failing Spanish.

I called the college and had an admissions expert help me register. When I hung up the phone, I thought, *I did it! I enrolled in college.* And guess what? I failed Spanish. Yep. But then I looked at the facts—I took Spanish alongside a massive biology class, and it was too much. Not to mention my negative mindset that set me up to believe I would struggle with it from the start. The next semester, I made a choice to try again. I signed up for four classes, including

Spanish. I made a conscious decision to put my full effort in and worked on thinking positively. That semester, not only did I pass, but I also actually fell in love with my Spanish class! I had created a fear about taking Spanish back when I was in high school and allowed it to hold me back for twenty years! Now, I had overcome that fear by changing my untruth into a success.

I went FSIR and:

- Decided to feel I was worth it.
- Let go of the past.
- Searched my heart and knew my purpose was to have my degree.
- Broke out of my comfort zone.
- Stopped letting fear hold me back.
- Stopped believing the untruth that I was average.
- Remembered how good I am at using a spoon and decided to put the time into learning something new.
- Knew what I wanted, and I went for it.

Keep in mind, there is not one person on earth who hasn't failed at some point and succeeded at another. Failure is nothing to be ashamed of. Never be held back by failure because every failure leads to success. Be inspired by it! I'm proud of you for failing! It means you're still trying, and you haven't given up. Failing is a wonderful part of life! If you don't accomplish your dreams today, wake up and go for them again tomorrow. Rise above the feeling of having to accomplish something to feel a certain way about yourself. Rise above the feeling that you have to punish yourself because you've failed in your past. Failing and succeeding are not what you think they are. Your life is not about wrong and right, win or lose, failure or success. Just do whatever makes you come alive and know that you might fail. But if you keep going at your dreams, success will come.

Life is a series of repetitive successes and failures. Use your life to love, learn, grow, and develop, and from that, you will find your purpose. Each day is an opportunity to create an honest relationship with yourself right where you are. It's ok to love yourself just the way you are. Make peace with yourself. Spend time with yourself. Listen to yourself. A great way to find out what you lack is to ask yourself what do you want from your life that you don't currently have? If you don't love your job, what do you wish you could do? Stop holding yourself back and remember you are worth it. You can now go for it! Throw away the old meaning of success and failure, be proud of yourself, and move forward.

If I send this book to a publisher and get turned down, I'm not a failure. Even if I send this book to thirty publishers and get turned down by all of them, that's still not failure. Failure only occurs when you give up completely. I know there are people who will value my book and will find encouragement in the pages between the covers, and that's why I refuse to give up. Success and failure happen every day, but neither one is the end of the road. Grow your self-worth, figure out your why, and keep pressing ahead.

- After decades of holding myself back, I discovered the power of believing in myself—I passed Spanish I, and then, I passed Spanish II!

Dr. Wayne Dyer believed that success and failure are unhelpful judgments against us and that instead; we should focus on our purpose. According to the world's definitions, success generally signifies the completion of something, while failure signifies the end of something. This gives both words a negativity that can hold you back. The world's definition of success implies that you are only worthy *after* you accomplish something, and failure means that your value goes away when you don't succeed. It is my hope that you never look at the world's definition of *success* and *failure* the same after reading this chapter.

I hope you know that your value is unconditional. Think of the hundred-dollar bill. It doesn't have to do anything to earn that value; it's worth one hundred dollars. You are the same—you are who you are. Your only job is to be you, so believe in your value. With self-worth comes the ability to love yourself. When you have these, the next thing that results will be discovering your purpose. I call them the trinity. In an ideal world, you would possess an abundance of all three. Most people only have a part of the trinity.

Consider the many successful people who are gone from the world today, like Kate Spade and Anthony Bourdain, Robin Williams and Heath Ledger, Stephen "tWitch" Boss (my favorite dancer), and the music world's Jarad "Juice WRLD" Higgins, Chris Cornell, and Chester Bennington. I didn't know any of them personally, nor do I claim to know the exact reason they're gone, but drug abuse, depression, and feeling empty are noted in their lives and causes of death. Each one of them had a part of the trinity. They knew their purpose, but somewhere along the way, they lost their self-worth or self-love to be able to be filled internally. They had everything most people only dream of and were considered the definition of success, and yet, they're *gone*. Proof that money, fame, talent, and material things do not fill your heart. Even finding and living your purpose is not enough! Success is not about an abundance of material items. It's about feeling fulfilled, enriched, and genuinely happy. Without nurturing the inside, no amount of fame or fortune will create self-worth or self-love. You must actively work to increase all three parts of the trinity—self-worth, self-love, and your purpose—to find true success.

There are people who stop going toward their dreams because they've failed, and there are people who hold themselves back because they haven't succeeded yet. You only have a certain number of days on this earth. No one knows how many. Each day you wake up is the day you have. So, each day, the goal should be to value yourself, love yourself, and go toward your purpose to live the life of your dreams.

Judging yourself by the world's definition of success and failure means your self-worth is low. Its why people stay stuck in loveless marriages or keep going to jobs they don't love. Most people believe they must be successful to value or love themselves, but only you can define the word successful for yourself. Always remember that your self-worth and self-love are unconditional. You must love yourself regardless of what you have done or not done. Don't forget the power of *yet*.

If you are not where you want to be or feel you haven't completed your goals yet, there is nothing wrong with that; but don't let it stop your progress. Keep going. As long as you're breathing, you can keep pursuing your version of success. If you've got goals to crush, crush them! When you're done with those goals, set more goals and then crush those ones. There is no finish line. In fact, I hope you never stop setting goals. I'm not done completing mine because as soon as I achieve one goal, I set another. I hope we are both still crushing goals when we're 100 years old!

Keep in mind, there is not one person on earth who hasn't failed at some point and succeeded at another. Failure is nothing to be ashamed of. Never be held back by failure because every failure leads to success. Be inspired by it! I'm proud of you for failing! It means you're still trying, and you haven't given up. Failing is a wonderful part of life! If you don't accomplish your dreams today, wake up and go for them again tomorrow. Rise above the feeling of having to accomplish something to feel a certain way about yourself. Rise above the feeling that you must punish yourself because you've failed in your past. Failing and succeeding are not what you think they are. Your life is not about wrong and right, win or lose, failure or success. Just do whatever makes you come alive and know that you might fail. But if you keep going at your dreams, success will come.

Life is a series of repetitive successes and failures. Use your life to love, learn, grow, and develop, and from that, you will find your purpose. Each day is an opportunity to create an honest relationship with yourself right where you are. It's ok to love yourself just the way you are. Make peace with yourself. Spend time with yourself. Listen to yourself. A great way to find out what you lack is to ask yourself what do you want from your life that you don't currently have? If you don't love your job, what do you wish you could do? Stop holding yourself back and remember you are worth it. You can now go for it! Throw away the old meaning of success and failure, be proud of yourself, and move forward.

If I send this book to a publisher and get turned down, I'm not a failure. Even if I send this book to thirty publishers and get turned down by all of them, that's still not failure. Failure only occurs when you give up completely. I know there are people who will value my book and will find encouragement in the pages between the covers, and that's why I refuse to give up. Success and failure happen every day, but neither one is the end of the road. Grow your self-worth, figure out your why, and keep pressing ahead.

Life isn't about success. It's about going further than you went yesterday. I hope you never become so successful that you feel there is nothing left to accomplish. Stagnation means death—death of joy, death of passion, death of purpose. Make new plans and try new things. If you don't succeed, so what? Get up and go for it again tomorrow. Or take a break and come back stronger and ready to try again. Focus on who you were born to be and do that.

So, what's holding you back? What do you fear that you are meant to fall in love with? Like many people, I feared failure. I want you to know that you can push past the fear of the unknown. On the other side might be a love you've never dreamed of, just like I discovered my love for Spanish. Go FSIR and listen to your heart's desire. Let it guide you toward your future. Forget success and failure. Just focus on your purpose and learn everything you want to learn. Fear is an illusion created by you to keep you safely tucked away in your comfort zone, but opportunity can't knock when there is no door.

So, are you a failure or a success? Nope. You're neither. You can be anything you want to be. Everything is possible because you are limitless.

CHAPTER 6

TAKE RESPONSIBILITY

There will be two kinds of people on your road to change, those who are proud of you and those who doubt you. Those who doubt you are a wonderful test to see how badly you desire change. These are the people who will, in the midst of you trying so hard to achieve your goal, quickly remind you of the "old" you and make your goals feel unattainable. While it might seem easier to quit and blame someone else, in the end, no one is to blame but yourself. Every choice is yours. No one can make you do or feel anything unless you decide. It's your responsibility to decide to make a change, and it's your responsibility to follow through. Be patient with yourself. Although this is a big responsibility, it's also simple. It's big because you are the only one responsible for what's going on in your life. It's simple because if your life isn't what you want it to be, all you have to do is make a change. What parts of your life are you not happy with? Whatever it is, change it. It's no one else's fault—the responsibility is yours alone. Happiness comes from within, but so does unhappiness. Go FSIR and ask yourself what you are unhappy about and what it would take to make you happy. Be honest. Otherwise, you're only lying to yourself.

In the space provided below, write down what you are unhappy with and truly want to change.

_____

_____

_____

_____

_____

_____

_____

_____

_____

_____

_____

_____

Why haven't you gone for it yet? Are you waiting for the time to be right? There's no time like the present! If the goal seems daunting, just remember you are worth it and take it one day at a time. If you never take that first step, you will remain safely tucked away in your comfort zone, getting nowhere fast. Or is the problem that you're blaming someone else? When you claim someone *makes* you feel a certain way, you aren't taking responsibility for your feelings. Blaming someone else is the easy road. It means nothing is your fault. By never taking responsibility for where you are (or where you're not), you will become stuck, unable to get to where you want to go. Blaming others is another shortcut to nowhere. Don't allow the outside world to have an influence on you. Take back your power to be able to change whatever you are unhappy about by acknowledging that it's no one else's decision but yours. No one *makes* you do anything or feel anything. You alone are responsible for your own happiness. All of your actions and your feelings are your responsibility, so *be responsible.*

The main reason you don't make a change or go for what you want is how you feel about yourself. If you were brimming over with inner worth, your outside world would brim over too. Staying where you are or making the choice to break out of your comfort zone is up to you. Your self-worth is up to you too. If you don't feel good inside and you think surrounding yourself with certain people, or having a relationship will change that; you're right. It will change that. But what happens when you are alone? What happens when your friends go home, or your relationship ends. What then? Does your worth go away? It shouldn't. It is your responsibility to find your inner worth and realize that it doesn't come from outside of you. Your worth can only come from inside. You've set up all of these rules in your life that try to keep you from feeling uncomfortable. It is your responsibility to remove the rules you've set up to make life predictable and let life happen to you. Know that you are so worth it that you can get through anything life brings your way. You can go for that job, new career, degree, or learn anything you want. You can handle things when you feel uncomfortable because you are solid with yourself on the inside. You can't be shaken. No one on the outside can affect your inside. That is your responsibility.

THE CHOICE IS YOURS

Some unhappy people create a bunch of rules because they're unhappy. They become rigid and stuck in certain ways. You always have a choice, even if your choice is to stay stuck. Or you can take responsibility and make changes. Yes, you'll be in unknown territory. You might even have to put yourself first. This is not a bad thing. Focus on what you need and go for it. Do you realize how much time you waste giving to thoughts of other people when they are away from you? Your job should be to focus on yourself. The future is unknown. No one knows what is going to happen, but the choice is yours who (and what) you focus on. If you choose to focus on outside things, they will always change. If you focus on the inside, that is constant and always with you. Your presence never goes away. Everything outside of you goes away at some point and I don't mean death. Your spouse, or significant other will need to go to work, your friends eventually go home, all the things on things outside of yourself will leave you. What's left is where you need to find your comfort, your peace. The choice is yours who you focus on.

Do not allow your brain to talk yourself out of change because of the unknown. Life is unknown. Each day you wake up, you have no idea what's going to happen. You don't know if a deer will jump out in front of your vehicle on the way to work. You don't know if you will win the lottery this evening. You don't know if someone you love is going to die—boy, do I know all about this one. No matter what happens in life, you'll get through it if you take responsibility, value yourself, and make the necessary changes to realize your happiness. The decision to make a change is no scarier than these unknowns—plus, you have more control over the conscious choices you make.

When my husband died, I was devastated. I had no control over the fact that he chose to be with another woman or the subsequent car wreck that took his life. What I did have control over was how I responded. Taking responsibility for your feelings is very easy to do once you're used to doing it, but it takes time to make this a routine part of your life. Any apprehension you feel is just your brain afraid of giving up your right to blame someone else for where you are, but wherever you are is only because of you. Your discomfort does not come from someone else; it comes from the off-balance part inside of you. When you are balanced, you won't be knocked off kilter when someone leaves. Instead of seeing it as giving anything up, view it as taking back your power—the power that allows you to center your focus inward which will improve, enhance, and change your life.

Admittedly, taking responsibility for my feelings after Shannon's death was a challenge at first because I wasn't used to it. I remember thinking, *how could he replace me? How could he drive drunk? How could he risk his life and leave our sons? How could he do this to us?* It was easy to think that way, but I also knew it would not help me move forward. Going FSIR saved me. I was able to use my past to decide to make the best of my present. Every time I got mad or sad, I decided to change my focus. I chose to focus on what I was thankful for, which were our two sons. By valuing myself and taking responsibility for where I was, I was able to do the best I could for my sons, setting an example and being a reliable presence for them.

I also recognized that my feelings were my choice, so I made a choice to not be a victim. If I held on to all the hurt, I would have stayed stuck forever. Instead, I focused on moving on with strength and grace. I decided not to dwell on the fact that my husband died with another woman beside him. I chose to forgive him to free myself to move forward with my life. Taking responsibility for my own life was the best decision I ever made.

You may not be able to control what happens, but the choice will always be yours as to how to respond. Are there things you don't want to accept, are not dealing with, or need to take responsibility for? Be careful not to set up roadblocks that prevent you from achieving your own progress and happiness. Roadblocks mean you are not willing to budge, and they speak volumes about how low your self-worth and self-love are. When you take responsibility for how you feel, every emotion becomes less complicated. You have the power to acknowledge it and the power to change your focus away from it.

Let's say you've been in a relationship for many years and you're unhappy. You don't like where you live, you're waiting for a ring, your significant other is too busy for you, or you're waiting for your partner to make you happy. If this is the case, I can help immediately. You're in a *relationship*. That means two people living two lives at the same time. Each one of you is responsible for your own happiness. Stop being avoidant and have an honest conversation. Both of you need to clearly tell each other what your needs are. The initial conversation should have happened before you began your relationship with this person, but occasionally, your desires shift or develop as the relationship does. Remembering to regularly have these discussions with your significant other is the best way to maintain that sense of open communication and clarity about your desires and expectations. You might have to realize that you've chosen your significant other because you were filling up your need to be needed by someone. They might not even be who you really need or want! That is your responsibility to figure it out too. Your self-worth cannot come from anyone but you. Your happiness cannot come from someone else. It's unfair to expect someone else to give you a sense of worth. If you're relying on someone else you are guaranteed to be disappointed, and it's unfair to the other person. Not only because it's a lot of responsibility to bear but they need to focus on what makes *them* happy. You need to know what you want and look for someone who wants many of the same things. It's okay to have separate hobbies and interests, but the big things—marriage, children, faith, finances, etc.—need to have a common ground between you and your partner to work for the long term. Being committed to your marriage and staying when things get rough is very powerful; living a life together that makes both people miserable is not.

Stop blaming anyone for how you feel or the choices you make. Remember, no one can make you feel anything. My husband didn't choose another woman to hurt me; he did it to make himself feel better. But even if he had done it to hurt me, it's still up to me how I choose to feel about it. The amount of self-worth and self-love you have will determine what you do when bad things happen. Sometimes things can go from bad to worse, but it's up to you to choose what you do with it—will you stay stuck or move forward? Keep working on building up your self-worth and self-love. The time has never been more perfect for you to find it on the inside. You will also find your purpose in there too. It's time to take responsibility for your own happiness. If you're unhappy, find out why. Go FSIR and find what *does* make you happy, then do it. No more excuses.

CHAPTER 7

YOU CHOOSE

In the early 2000s, Shannon and I were living in a small, remodeled house in Granton, Wisconsin. I loved everything about our home. It was right in town, and though it was small, so were our bills to live there. The size also made it easy to clean, especially with our amazing routine. Each Sunday morning, we would get up and make breakfast. After we ate, we'd turn the music up and clean the whole house together in about an hour. It was my favorite day of the week because I felt like the strongest team whenever we worked together to clean that little house.

It was during this same time that Shannon started his business as a home builder. Although I was happy with our small home, it was nothing like the stunning log homes he was building for our customers. Having grown up in a small trailer home, Shannon had always dreamed of one day living in a big home like the ones he built. There was no doubt that living in a grand house would make him feel successful and impress our customers when they came over to discuss their home plans.

When he found an opportunity to purchase property in Neillsville on the Black River, we decided to build a gorgeous log home of our own. I knew Shannon would have to work a lot more when we built the big house because our mortgage would be tripling. This meant the days spent cleaning the entire house together were over, but I knew this new house was very important to Shannon, so it was important to me.

We had made the decision to build the big house together. We both picked out everything for the new house. He picked a giant front door, we both chose the granite countertops, the paint colors, and everything we wanted in that house. Both of our paychecks went into our joint bank account and we both paid for the big house. We were a team. Although I preferred our smaller home in Granton, we had made every choice together.

Most unhappy people blame others for where they are or are not. Unhappy people rarely take responsibility for anything, believing life happens *to* them, not because of them. In this belief, they take on a victim role. According to the Oxford Learner's Dictionaries, a victim is "a person who has been attacked, injured, or killed **as the result** of a crime, a disease, an accident, etc." "As the result" is the key phrase in that definition. You may be a victim of the things that happen outside of your control, but you can't be a victim of the choices you make afterward. Being unhappy is not the result of anything except your choices. You have a choice to be where you are right now. You are not a victim of your life, but an active participant.

Who decides to wake up each morning and get ready for the job you go to? You. Who decides what you eat, drink, and wear? You. Who decided your vehicle? The roads you drive on? Who chooses every word you say? What do you share and what do you keep to yourself? You. When you go home this evening and make supper, you will make what you choose to make. Everything you do is because *you choose* to do it. If you are unhappy, it's because you're choosing everything in your life that's making you unhappy.

When Shannon died, I couldn't afford the mortgage for the big house on my own. I knew that when we built the house. As I scrambled to figure out a solution, people asked if I regretted marrying him because of how it ended. The answer was a no-brainer: no, of course not. Not once did I wish I hadn't fallen in love, gotten married, or moved to Wisconsin. We both *chose* each other, and for a time, life was good. The only detail I am a victim of is that neither one of us thought Shannon would die at thirty-nine years old, but everything else was and is a choice. Realizing that is so important.

Most unhappy people forget they chose every aspect of their lives. If you're unhappy, choose to make a change. Maybe that change is simply what you choose to focus on. Even though our marriage ended poorly, I choose to focus on the fact that I got my two wonderful sons out of the deal. I choose to be strong and gracefully move forward with my life. I choose to continue to seek new knowledge, skills, and opportunities so that I continue to grow until my dying day.

Choose to be an active participant in your own life. To cure unhappiness, go FSIR and change how you feel about yourself. If you're unhappy with others, ask yourself why and do something about it. If you've been unhappy for years, it may be time to stop doing the things that are keeping you unhappy. Nothing is out of your control. If you are still unhappy or struggle to make any lasting changes, it's likely due to how you feel about yourself. Find self-love and self-worth and remember who you are. You are worthy. You are amazing. There is only one you, so connect with yourself and begin the journey toward what makes you truly happy.

It's not easy to realize that your choices may have resulted in you being unhappy. Changing the way you think and feel about yourself can be scary. Taking responsibility might also be new. Today might even be the first time you realize that *your choices* are the reason for your unhappiness. Focus on the fact that this means you can also make the choice to pursue happiness.

I'm not asking you to jump in and change everything all at once. Start by building up your self-worth and self-love, because feeling good on the inside makes everything on the outside possible. Make a pros-and-cons list and prioritize the changes you'd like to make. Then make a one-year plan and begin. Time and patience are required for such big decisions, but remember, this is your life—you choose. Stop fighting against yourself and start fighting for yourself.

If you are a victim of domestic violence, sexual assault, or thinking of suicide, please seek help. There are people who can and want to help you. Please call immediately.

National Domestic Violence Hotline: 800-799-7233
Suicide & Crisis Lifeline: 988
National Sexual Assault Hotline: 800-656-4673

CHAPTER 8

THE IDEAL SELF

I t's time to go FSIR one more time and create your ideal version of yourself. I'd like to publish two more books besides this one. I'd also like to live my dream career as a motivational speaker where I travel around the US and inspire high school students, businesses, youth groups, and audiences of all ages. I'd like to successfully complete Jose Silva's mind control course and be well on my way to skiing every mountain in North America. I also dream about being happily married and traveling the country in our van and airplane, living the life I dream of as an inspiration to myself and anyone I meet.

Your turn. To find the ideal you, think back to Chapter 1 when you wrote who you are now—that's your baseline. Next, think about what makes you come alive on the inside—that's your purpose. You know, the thing(s) you'd do if you had all the time, money, and knowledge in the world. Then, listen to your inner-self and admit everything you would like to do or become—this is your ideal-self waiting to be unleashed.

Remember not to think about this negatively. Your ideal self is not something you should ever hold against yourself. Rather, use it to motivate and encourage yourself to grow. Your ideal self is an embodiment of your goals and who you are working to become. All the things you've learned so far are meant to increase your worth and allow you to go toward the life you dream of. All the work you've done so far in this book should give you tools you didn't have before and make life easier. This entire process is meant to unleash your potential, crush your fears into oblivion, and allow you to become who you've always dreamed of becoming. So, let's think about some of the changes you want to make.

What things are you doing currently that you don't want to do? What do you want to be doing instead? Why aren't you? What scares you about making these changes? If there were no negative consequences and money, time, age, location, and education did not matter, what would your ideal self look like?

_____

_____

_____

_____

_____

_____

_____

_____

_____

_____

_____

_____

_____

_____

_____

_____

_____

I'm excited about your ideal self! I'm excited about mine too! But if you never try, you will never get any closer to your goal. Know that you are worth it and anything you've ever wanted is possible. It's up to you to make your dreams a reality. Once you reach your first goal, set another, and reach again.

I wish someone would have taught me how to go toward my ideal self sooner. If I had known how to find self-love, self-worth, and my purpose at a younger age, I could have broken out of my comfort zone and written this book a long time ago. What could you have done sooner? The good news is it's never too late to start! Your ideal self is waiting for your much-needed attention. No matter where you came from or what your past was like, you are a good person and a worthy person. You deserve happiness. You deserve to love who you are. You are here for a reason, and you are exactly where you should be. Everything you need to become the ideal you, is already within.

CHAPTER 9

NEVER GIVE UP

The last tool I want to give you is the courage and wisdom to never give up. As long as you keep trying, you *will* accomplish anything you set your mind to. It doesn't matter how long it takes and it doesn't matter how many times you must start over. If you never quit, you will always finish. The key is knowing that anything you focus on will win. I could have written this book a long time ago. I allowed my brain to remind me that there are more qualified, more educated, and more successful people out there who were also writing books. And my brain was right. There were people writing books who were better qualified, smarter, and more educated than I am. But there's also people out there who are less educated, less qualified, and who haven't been through what I've been through, and you know what they're doing? Yep. They're writing and selling their books. You can tell yourself why your dream won't work, or you can go FSIR and use the tools in this book to put your energy into achieving your dreams. Never give up. No day spent going toward your dreams is ever wasted, even if you haven't succeeded *yet*. And no, it's never too late to begin.

Remember your ideal self from the previous chapter? We're going to use "I am" statements to rephrase your ideal self-statement, as though what you desire has already come to fruition. Here's mine:

I am a published author of three books. I am a motivational speaker. I am a traveler of this earth. I am full of self-worth. I am living the life I dream of. I am happy. I am worthy. I am loved.

What about you? What are you going to change? What are you going to make happen? Think about all the things you are going to accomplish in your life, no matter what they cost, how much time they will take, or the effort they will require.

Write about your ideal self, using "I am" statements.

I am _____

_____

_____

_____

_____

_____

_____

_____

_____

_____

_____

_____

_____

_____

_____

_____

_____

_____

Guess what? Each one of those things you wrote _will_ happen if you don't quit.

Happiness can be realized when you:

- Know who you are now and discover your purpose.
- Use your past to improve your present.
- Get out of autopilot and connect with yourself.
- Meet your brain and learn to conquer your fears.
- Remember spoons and try something new.
- Disbelieve your untruths and seek healing.
- Understand your triggers and realize the power of a positive mind.
- Forget success and failure and believe in yourself.
- Take responsibility and remember that you get to choose.
- Create your ideal self and never give up.

May you never feel unworthy, unwanted, unusual, uninspired, or unloved again. You don't lack anything. Use your new understanding to make your life unlimited. I found happiness when I realized my worth and love is inside of me. Yours is too. Once you fully understand all the things above, you won't ever feel the need to look for your happiness on the outside again. No one else holds the key to your happiness. You've been the keeper of the key your whole life. You only need to reach out and unlock the door.

I hope every thought you have from now on changes your life for the better. I hope every dream you have comes true and your heart opens to a whole new world—one you never believed was possible. I hope the emptiest places in your soul are filled with your new self-worth, self-love, and driving passion to live your purpose. I hope you feel limitless. I hope you find your happiness realized.

I'd love to hear from you! Visit my website at www.HappyAmy.net or send me an email at happyamy75@icloud.com.